Architectures for Enterprise Integration

Architectures for Enterprise Integration

Edited by

Peter Bernus
Griffith University
Queensland
Australia

Laszlo Nemes
CSIRO
Victoria
Australia

and

Theodore J. Williams
Purdue University
Indiana
USA

Published by Chapman & Hall on behalf of IFIP and IFAC

IFAC

SPRINGER-SCIENCE+BUSINESS MEDIA, B.V.

First edition 1996

Originally published by Chapman & Hall in 1996
Softcover reprint of the hardcover 1st edition 1996

ISBN 978-1-4757-4541-2 ISBN 978-0-387-34941-1 (eBook)
DOI 10.1007/978-0-387-34941-1

A catalogue record for this book is available from the British Library

Printed on permanent acid-free text paper, manufactured in accordance with ANSI/NISO Z39.48-1992 and ANSI/NISO Z39.48-1984 (Permanence of Paper).

Contents

List of contributors

Peter Bernus (Griffith University, Australia)
David Chen (Bordeaux University I, France)
Guy Doumeingts (Bordeaux University, France)
Hong Li (Purdue University, USA)
Laszlo Nemes (CSIRO, Australia)
John P Shewchuk (Virginia Polytechnic, USA)
Jakob Vlietstra (AT&T Networks International, the Netherlands)
Theodore J. Williams (Purdue University, USA)
Dick Zoetekouw (AT&T Networks International, the Netherlands)

Preface

This book is the result of the joint efforts of many people who have worked together over the past three years (1990-1993). A group of manufacturing engineers, computer scientists and information technology managers formed a Task Force in 1990 to study, compare and evaluate the different available architectures for enterprise integration which were in the open literature.

This joint working group was formed under the auspices and sponsorship of the two major international federations on automation and on information processing. These federations were the International Federation of Automatic Control (IFAC) for automation and the International Federation for Information Processing (IFIP) for the latter field. The basic task assigned to the Task Force was to study the available architectures and make recommendations for the future identifying one existing architecture as best. Failing to do this, the second best option seemed to be to build on an existing system by supplying the apparently missing capabilities. It was recognized that as the last resort it might be necessary to specify a whole new architecture from the characteristics of several of the others. The Task Force reported its findings in the predecessor report to this book.

A summary of the report of the Task Force was presented at three separate major conferences of IFAC and IFIP during the second half of 1993. From the response generated there it was felt that the contents of the detailed study report would be of interest to many people, hence this book. Industry could find this book useful when determining the best methods to be used to help integrate their manufacturing operations. Researchers both in industry and academia should find it a valuable source of reference when investigating new areas of research and when comparing their present or proposed activities with the current state-of-the-art.

Although is not a textbook as such it does, hopefully, contain enough information for the reader to understand the main principles and methodologies of the integration architectures described. Thus, the book should also find its way to the desks of university staff and students.

Integration is, and will continue to be, a key factor in building and maintaining comprehensive enterprises. What we provide here is the state-of-the-art of the methodologies required for use by everybody who is interested in expanding his/her knowledge in that field.

We would like to encourage all readers to report back any omissions or errors in the book.[1] We also welcome suggestions on how to improve further editions of this book when we have occasion to add our new results to the existing material.

A complete list of the membership of the Task Force is given in Appendix B.

The Editors

1. See http://www.cit.gu.edu.au/research/groups/ei/erratum.aei.html on the World Wide Web.

Acknowledgements

Special contributions to the work reported herein have been made by the following persons:

Professor Ted Williams, chairman of the Task Force, and of the Purdue Industry Consortium that worked out the PERA Reference Architecture, as well as the associated Implementation Procedures Manual, developed many of the ideas including leading the one-on-one mapping of reference architectures, with the help of John Shewchuk and Hong Li, and Dick Zoetekouw (who initially proposed the one-on-one mapping).

Professor Guy Doumeingts, together with David Chen prepared the state of the art survey on models, architectures and methods for integrated systems design as included in this book; special mention should be made of Bruno Vallespir for his valuable insight in discussions.

Dr Peter Bernus, vice-chair of the Task Force, initially proposed and developed the detailed questionnaire for the evaluation of reference architectures and analysed the results. He also developed the matrix presentation of reference architectures and prepared with Ted Williams, as well as the help of Francois Vernadat and David Chen, the detailed mapping of the three major architectures to the evaluation matrix.

Dr Laszlo Nemes, together with Peter Bernus worked on the development of the detailed questionnaire evaluation form and the matrix presentation form of reference architectures. He also chaired the Canadian section (Toronto) meeting of the Task Force and has given his valuable management experience to the Task Force's strategic direction. He was also chairman of the IFAC Technical Committee on Manufacturing Technology (MANTEC) from where the Task Force originated.

Ing Jakob Vlietstra, in the period of the first triennium of the Task Force, was information technology director of AT&T Networks International, the Netherlands (formerly AT&T Phillips), chairman and initiator of the AMICE consortium which developed CIMOSA. He has contributed to this book by his detailed explanations of the technical as well as strategic aims of the CIMOSA work. He also proposed the short form evaluation questionnaire.

Jim Nevins was the initiator, in the IFAC Technical Committee on Manufacturing Technology (MANTEC), of the Task Force, and had a major influence on the establishment of the Task Force's scope and working plan at the initial Bordeaux meeting. The Task Force Members express their sincere gratitude to Mrs. Judy Ruzich, Mrs. Barbara Beaver and Ms. Paula K. Slyder of Purdue University, for their speedy and accurate compilation of this text and its earlier report and the numerous figures involved from their original massive, hand-written draft form. We also want to thank Dr Mingwei Zhou of CSIRO for his help in the proofreading process.

1

Introduction

L. Nemes and P. Bernus

1.1 ENTERPRISE INTEGRATION: THE NEXT GENERATION OF MANUFACTURING SYSTEMS AND THE NEW MANUFACTURING PARADIGM

1.1.1 Custom made products

The much touted trend of producing custom made products in the quality and with the efficiency of mass production has by today a foothold in almost every part of industry. Two major effects help this trend to gain ground: the customer's desire to get products in the exact specification desired, and the manufacturing enterprises desire to produce goods only for existing customers, rather than for would-be ones. The above two requirements underly competition because once a need is identified competitive forces will dictate to react quickly and professionally, flexibly adapting to the market need. For this reason agility and flexibility are two important properties companies strive at.

1.1.2 Virtual and extended enterprises

The traditional view of an enterprise is a hierarchical organisation. In big companies with diversified business profiles an enterprise may have meant that a branch office or a plant operated as an independent business unit.

Today very few companies design and manufacture all components in house and the view of the enterprise as an independent unit is not very adequate. In fact most of the agile enterprises cooperate with a large number of suppliers and sub contractors. Components and subassemblies are therefore manufactured outside the premises of the product vendor. More recently they are being designed in close cooperation with partners in the supply chain. Some suppliers are in strategic relations with a number of product vendors.

Companies must focus on the business processes that encompass the entire chain of value adding activities, called the supply chain, because it is this complete process that determines the overall result, not the process within a particular company 'A' alone. In this sense a company 'A' will have business processes of which a part are being performed outside the physical boundaries of the company. For this reason a company 'C' may be a part of the core business of company 'B' by delivering some components or subassemblies to it. The business decision, 'make or buy' or 'make or have it made' is certainly a key factor when defining an extended, integrated enterprise. Business activities (business focus) may change quickly and so should change the respective business process. The dynamism, the time dimension, of the enterprise design and implementation throughout the whole life-cycle of the enterprise is critical, and will be even more critical in the future. Virtual or extended enterprises will be formed and dissolved as quickly as the underlying business processes are changed [1]. Computer-aided tools therefore should be able to cope with the highly distributed cooperative nature of the virtual, extended enterprise concept and help companies design and redesign enterprises as the need arises.

1.2 SYSTEM BUILDING PRINCIPLES

The traditional concept of computer integrated manufacturing aimed at providing computer assistance for various elements of the product development, manufacturing and management processes. Integration in this context meant to form the interfaces and implement the information links among computer aided activities to allow direct and timely cooperation among them. Consequently one important aspect of integration has become the provision of interoperable information systems [2] that allow easy coupling between distributed elements of the material and information processing system, that is, the enterprise.

Although the above endeavour is still very timely it must be foreseen that the mere possibility of ubiquitous information technology will enable a more profound change, allowing companies to step beyond integrating their applications. Enterprise integration is the step beyond that: it asks questions about the core business processes in a organisation and concentrates on how to perform these tasks better, whatever modifications are deemed necessary to achieve this goal. Furthermore Enterprise Integration considers the process that leads to better solutions, not only the better solutions themselves.

In such a context Enterprise Integration captures and describes processes, strategies, organizational structures, resources, goals and constraints of the enterprise. It specifies the business process requirements, identifies the solution options, presents alternative designs and provides implementation paths at strategic, tactical and operational levels.

The entire business process is studied and described in terms of three inter linking networks:

1. Material flow and transformation together with service provision (i.e. the process which fulfills customer needs).
2. Information flow and transformation (knowledge and information processing in any form) that is needed to manage and control the enterprise.
3. Supporting organizational structure of people who carry out the nonautomated parts of the tasks in the two above mentioned processes. This component is essential, since people make decisions based on experience and incomplete information.

What are the best structures of business processes, organisation, and management is of course open to separate investigation: the overall method of engineering enterprises should rely on the results of the disciplines of organizational design, facility design, management science, information systems and communication. Several recently emerging research areas address the issues of which structures are suitable for efficient high quality systems design.

The ideas behind holonic manufacturing, fractal factory, agent based integration and coordination theory all address the same issue (with slightly different focus): how is it possible to design a complex enterprise, that acts as an autonomous 'agent' following its own plans to achieve its objectives while cooperating with other agents for certain common objectives? Is there a simple principle of building such an 'aware' enterprise out of participating agents? What are the means of building, coordinating, planning and operating these lower level agents (which themselves may be composed of similar, lower level agents)? Answers to these questions may dramatically reduce the complexity of today's enterprises by having to consider crucial management questions only at one level at a time.

1.3 METHODOLOGIES AND THE ENTERPRISE AS A PRODUCT

In order to manage the change process of the business effectively, enterprise integration methodologies are needed. Methodologies exist in the framework of an architecture, and rely on modelling techniques, languages, and implementation know-how (such as previous knowledge of typical successful management-, process-, and organizational structures).

It is probably less well known that the presently widely used term of 'business process reengineering,' often referred to by consultants and understood by company executives is just one, albeit important, aspect of Enterprise Integration (EI) [3]. BPR, as it is abbreviated, focuses on the review and redesign of the core business process of an enterprise and it uses either common sense practice or professional tools offered by the EI community. There is a close cooperation between researchers of BPR and EI. What BPR does not address at this stage is the need to continuously engineer the business process, instead of just reengineering it once.

With the advent of globalization of economies, enterprises are more and more viewed as products themselves. This is quite obvious for green field companies as they are designed, built and put into operation as the customer requires. The 'product view,' however, is also true for already existing enterprises: new business priorities have to be analysed, changes have to be specified, designed and carried out. The design of an enterprise may take a long time and involve many people, therefore methodologies with a phased approach are preferred; i.e. methodologies in which the overall design can be split into implementation projects easily according to the benefit expected from each such project.

Business, manufacturing practice, organizational structure, information infrastructure keep changing to provide the most effective framework for the core business of an agile enterprise. Therefore enterprises should often be redesigned, and these modification could be significant. This creates the need for methodologies which include the element of constant renewal on a needs basis (should the need be based on internal or external factors).

To design enterprises and manage them through their entire life-cycle we need fundamental principles and powerful tools that support methodologies. An important challenge is to actually use the models of the enterprise to enact the business processes rather than keep the business process models on the shelf, since in this latter case there is little chance for the enterprise to keep an up to date picture of itself and much of the effort initially invested in modelling the processes in the enterprise is lost as reality diverges from those model. Builders of machines, electrical devices, or computer software have been using design and analysis tools for a long time and manage to keep their product models up to date and reusable, in fact these models are the basis for the entire product life-cycle. These tools, however, are product oriented and specialized to the type of product at hand. The enterprise as a product needs the matching enterprise engineering tool, which is the basis of the enterprise's implementation and constant renewal, therefore tools needed for enterprise integration are not just more complex ones but substantially different from tools used for designing conventional products.

1.4 MAJOR PROJECTS

There are a number of large scale projects addressing the future of manufacturing industries. Here we only deal with those international efforts which aim to pool resources and skills to address major and common issues. More precisely we only refer to those which have relevance, or which even require the tools and methodologies offered by Enterprise integration architectures.

1.4.1 The Intelligent Manufacturing Systems (IMS) programme

The program was initiated by Japan in 1989 and was aimed to establish a $5 billion international research program for the next 5 years. The program was established after two years of negotiations and the completion of an international feasibility study. The identified technical themes were selected to address the main driving forces of modern manufacturing.

1.4.1.1 The driving forces

(a) Changing markets
There will be revolutionary developments which will profoundly change the manufacturing paradigm to base production on smaller units of higher quality. Advanced countries will shift from the present *make-to-stock*, sell-from-stock practice to manufacturing only the products wanted in only the quantities wanted. Companies will manufacture to actual orders only. The change in market will probably be drastic in advanced economies, from traditional growth markets to fragmented markets. Production will move closer to the consumer, since distribution cost is likely to become increasingly important.

(b) Customer orientation (flexibility)
The enormous increase in production variety and the great reduction in volume of a single product will affect all producers. Customers will have more individualistic desires. They will be interested in the details of products and services. Consumers will participate in the design and production by selecting a combination of safe and compatible components via electronic communication. This will unite *producers* and *consumers* again into '*prosumers*.' The prosumer will not see the model in a showroom since it

will be specially manufactured for her/him. The new model ideas will be viewed as electronic images. This practice will be quite common not only in the car industry, but in every manufacturing sector.

(c) Environmental requirements
The new generation of the consumer society demands that production processes should be environmentally friendly. The process should not discharge pollution to air or water. It should recirculate all material necessary for the manufacturing processes,

Design and manufacturing processes should ensure that all products and product materials should be re-used in order to minimize land fill waste.

(d) Human and social requirements
The hierarchical management structures provide clear separation of responsibilities, but different metrics are used to evaluate the performance of each function. In order to increase performance, management will instead establish a multi-functional project team whose coordination goes well beyond the traditional reporting duties. The project management team should play multiple roles to steer the project towards its vision, coordinating opinions from different functions throughout the company, stimulating innovative ideas and offering new knowledge for project members.

(e) New technologies and new production paradigms
Enterprises have already broadened their market beyond providing competitive products on competitive price. They offer products, services, technology, personnel, equipment, sites, and information (consulting, licensing, leasing, education). These trends will accelerate and will make competition more intense and difficult. Because customers will demand a wide range of products and services, rival companies must not only compete but also cooperate. It is more than likely that manufacturing will be regarded as a specialized form of *service*. This new manufacturing paradigm will integrate precision technology, information, communication technology and intellectual creativity.

1.4.1.2 Technical themes

The themes address five main areas for collaborations. The reader will note that four of them are closely related with the architectural issues. We only list the themes with their keywords as approved by the International Steering Board and refer the readers to seek further details from their regional IMS Secretariat.

- Virtual/extended enterprise issues
  - (a) Information exchange across the extended enterprise
  - (b) Team work

  (c) Architecture for support of engineering cooperation
  (d) Assignment of cost, risk and rewards
- Total product life-cycle
  (a) Models for future manufacturing systems
  (b) Environmental protection
  (c) Network Systems for information processes
  (d) Economic models
- Strategy/planning/design tools
  (a) Business process re-engineering
  (b) Analysis and development of manufacturing strategies
  (c) Planing in an extended virtual enterprise environment
- Human/organizational/social issues
  (a) Promotion of manufacturing as a discipline
  (b) Work force education and training
  (c) Organizational learning
  (d) Performance metrics
- Manufacturing processes
  (a) Clean manufacturing
  (b) Energy efficient manufacturing
  (c) Technology innovation
  (d) Flexibility

The global IMS program will be guided by an international IMS Steering Committee (ISC), with membership drawn from the participating regions. An interregional secretariat provides support for the IMS Steering Committee.

1.4.2 Technologies for business process re-engineering (a European Commisson project)

This project is part of the ESPRIT fourth framework. The mission of the project establishes the objectives 'To enable European businesses to compete more effectively within global markets by adopting and integrating best business practice in the areas of business process methodologies, information technology and human resource management.'

The documents put special emphasis on the enterprise integration issues. 'There will be a complementary activity in the integration of enterprise wide systems which will address the provision of tools, methodologies, and technology, filling strategic gaps as identified in conjunction with other activities. The research and technological development (RTD) activity will improve the potential for the next levels of business best practice.'

These two examples clearly show how important the Enterprise integrations architectures are. The authors hope that the work presented in this book will be of some value for those who are building integrated enterprises.

1.5 REFERENCES

[1] S L Goldman, R N Nagel and K Preiss. Agile Competitors and Virtual Organizations. Van Nostrand Reinhold Publishing Inc. 1995
[2] 'Interoperable Database Systems,' Proceedings of the IFIP WG 2.6 Working Conference DS-5, Lorne Vic. Aus, 1992, D.K. Hsiao, E.J.Neuhold and R.Sacks-Davis (Eds). Amsterdam, North-Holland, (1993) 347p.
[3] M Hammer and J Champy. Reengineering the corporation. Nicholas Braedly Publishing, London. 1993.

2

The concept of enterprise integration

T.J. Williams, P. Bernus and L. Nemes

2.1 THE PURPOSE ASSIGNED TO THE TASK FORCE

Managers and information scientists, control engineers and others have long been intrigued by the concept that one should be able to correlate and use, at any one time, all of the information then available concerning an enterprise, to systematically improve the operability, profitability, agility, competitivity (and many other important characteristics) of these systems.

Such a desire was, for example, the basis for the development of the field of computer integrated manufacturing which is built upon the ability of the computer to handle vast amounts of data or information in very short periods of time.

However, although many projects have been initiated in many types of industries in recent years to try to achieve these dreamed-of-benefits, the results achieved, in many cases, have been disappointing at best. In addition, where successes have been achieved, no clear path for obtaining future successes in subsequent applications has become clear.

The following is an important summary of the state-of-the-field of Enterprise Integration (including CIM), and points out what had been learned prior to the work of the Task Force as described herein.

2.2 THE STATUS OF ENTERPRISE INTEGRATION

An Integrated Manufacturing System solution cannot be bought off the shelf; each firm must be involved in devising its own, which explains why methodologies must be made available so that CIM systems can be built.

Designing a CIM system meets with a number of difficulties:

- The system is extremely complex, so some special techniques must be used to understand this complexity in order to efficiently act on it and to define the possible intervention area of every operator related to their skills.

- The system must take into account not only the technical points of view but also the economic, social, and human points of view in an integrated way.
- The knowledge which is necessary to design a system cannot be found in a single person; designing requires team work.
- The initial status of the system is not a matter of chance; this initial status must therefore be taken into account in order to well understand those specific constraints under which the system operates on the one hand and, on the other hand, to avoid the redesign of those parts of the system which are satisfactory. Because of today's economic situation, this point becomes more and more relevant; many industrial systems must evolve faster and faster but with as low a cost as possible.

The term 'methodology' as used here means a consistent set of components which are:

- A reference model globally and generically showing the structure of the project system to be studied.
- One or more modeling formalisms enabling the build up of the model in order to study and evaluate it.
- A structured approach for the overall program leading step by step from an existing system to a future system taking into account evolution objectives and specific constraints
- Performance evaluation criteria with which the system can be evaluated in relation to several points of view (economics, reliability, etc.).

Generally speaking, a structured approach or methodology is a set of steps to be followed to solve a problem. Within the framework of an integrated manufacturing system design methodology, the structured approach must cover all of the life-cycle of the integration project which is split into several successive stages (analysis, design, development, implementation, and operation). Every step of the methodology must be precisely defined and based on a standardized project structure giving the set of actors for which the work must be precisely defined as well.

Defining very precisely all steps which are required to go from concept to specification to implementation is essential. During each step some models are built and checked. The consistency of these models and results for each step must also be constantly checked [1].

As used in the context above, an architecture is any method (drawing, model, description, etc.) for giving the structure or framework showing the interrelationship of all of the parts and/or functions of a device, system or enterprise. A reference architecture is a collection of the overall generic functions, descriptions, or behaviors of many (hopefully all) types of systems and their associated structures or frameworks.

Thus an architecture, or preferably a reference architecture, should form the basis for the development of a device, system or project for carrying out an information integration program for a manufacturing or other enterprise. The problem remains, however, of: What is the best architecture for such purposes?, and How may it be developed? This then was the task assigned to the Task Force which developed the report contained herein.

The above statement characterizes very succinctly the industrial requirements which needed to be considered by the IFAC/IFIP Task Force on Architectures for Integrating Manufacturing Activities and Enterprises as it faced the task assigned to it by its parent bodies. Thus our task of evaluating manufacturing integration architectures must consider not only the 'what' of integration but also the 'how' of integration as well. That is, how integration is accomplished in terms of the development of the system, not just the overall design of the computer system (for example) needed to achieve integration.

This line of reasoning is reinforced when one reviews the field of integration architectures. An architecture, as just noted, can be defined as a description (model) of the structure of a physical or conceptual object or entity. Thus there are two and only two types of architectures which deal with the integration of manufacturing entities or enterprises. These are:

- The structural arrangement (design) of a physical system such as the computer control system part of an overall enterprise integration system.
- The structural arrangement (organization) of the development and implementation of a project or program such as a manufacturing or enterprise integration or other enterprise development program.

All of the reference architectures discussed in this book should therefore be of the second type in order to fulfill the requirements voiced above. (It is noted in passing that architectures of the first type can well become major subunits of architectures of the second type, for instance as examples or guides in the functional or detailed design phase of the integration program development).

There were only three major architectures in the open literature which were known to the Task Force as of this writing which fall into Category 2 above. These were:

- The Open System Architecture for Computer Integrated Manufacturing – CIMOSA, as developed by the European CIM Architecture Consortium (AMICE – a backward acronym) under ESPRIT Projects 688, 2422 and

5288 of the European Community. This work was initiated in 1984 (hereafter called CIMOSA). [3]

- GRAI-GIM. The GRAI Integrated Methodology as developed by the GRAI Laboratory of the University of Bordeaux in France. This work resulted from production management studies initiated at the GRAI Laboratory as early as 1974. It has taken its current form since about 1984 (hereafter called GRAI-GIM). [1,2]
- The Purdue Enterprise Reference Architecture and the related Purdue Methodology as developed at Purdue University as part of the work of the Industry – Purdue University Consortium for CIM. This latter work started formally in 1989 but bears on the Purdue Reference Model developed in 1986 and earlier work of the Purdue Laboratory for Applied Industrial Control dating back to the mid seventies (hereafter called Purdue or PERA). [4,5]

These then are the three architectures which were extensively examined by the Task Force and which are thoroughly described and evaluated in this book.

It must also be remembered that manufacturing integration is not automation and/or computer application per se, although computers and automation are usually involved in its implementation. Manufacturing or enterprise integration, on the one hand, is the collection, reduction, storage and use of data (and/or information) from the business entity involved, plus its environment, in order to optimize the operation of the business entity as a whole according to the criteria established by that business entity's management. On the other hand, it must also include the integration of the raw material, intermediate and final product flows, as well as machine organization, to further the enterprise optimization possible. In addition, the actions and influence of the human staff involved must also be thoroughly integrated with the manufacturing or other enterprise operations as well as the information integration carried out by the computers and related devices.

2.3 THE RELATIONSHIP OF THE TASK FORCE'S WORK TO THAT OF OTHERS

It is the purpose of this book to define enterprise reference architectures; present what they are and what they can be used for; treat their various forms; and project the future of the field of enterprise integration as it appears in the findings and recommendations developed by the Task Force in its investigation. The Task Force believes that it has developed new and important concepts in its work. These are presented herein. The Task Force further believes that these concepts, findings and conclusions may be helpful to others as follows:

Information technology experts, who develop and market comprehensive business information systems (software or hardware), information systems development methods, and software engineering or CASE tools, may find guidance on how to complete presently available but unfinished enterprise system architectures so that these architectures can become truly generic, and therefore applicable to the entire enterprise, and indeed to all types of enterprises regardless of the industry or type of endeavor involved.

The state of the art of enterprise reference architectures is presented here in order to allow the owners of proprietary architectures to assess their own products in view of this present description of the generic architecture requirements.

Developers of control systems (both hardware and software) will hopefully find that this book can also give them guidance in a way similar to that for the information technology experts. The control engineers situation differs in our view only in that they cover a different subpart of the entire enterprise domain from the information technologist. That is, they concentrate particularly on control system description techniques, methods for control system design, and/or tools to support these various methods rather than on data handling techniques, data base design, etc.

Developers and users of life-cycle methodologies in consulting firms, standards bodies, or large multinational companies may find our work useful to them to help them to evaluate their currently applied methodologies in the light of the analysis given herein. This should then help them to formulate their own future strategies to allow courses of action which would lead to a more complete enterprise integration methodology than those available today.

Policy makers and funding agencies may also find here a set of overviews and evaluations which they can then use for assessing ongoing research results. Our results may also be useful for deciding on the future research directions they may wish to pursue.

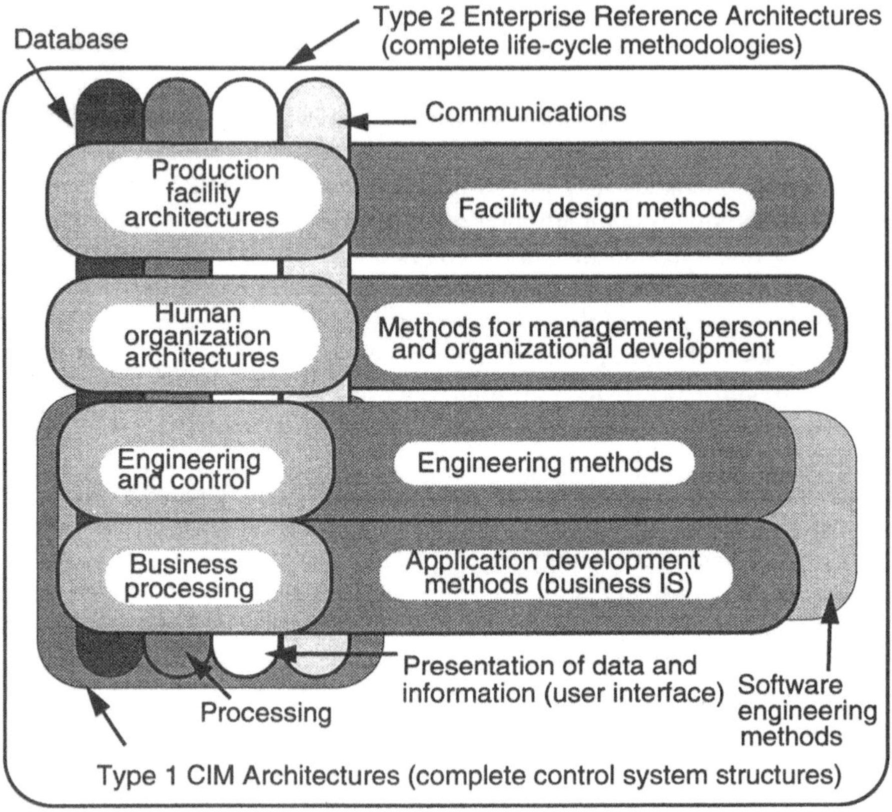

Figure 2.1 Representation of the separate classifications of architectures as used in this book.

2.4 A FURTHER DEFINITION OF THE RELATIONSHIP BETWEEN ARCHITECTURES

Fig. 2.1 presents a diagram which illustrates the way in which the Task Force has defined the relative scope of Type 1 and Type 2 Architectures. It will be seen that the Type 2, Enterprise Reference Architecture, comprises the whole context of the diagram of Fig. 2.1. That is, it presents a life-cycle development methodology for all aspects of the enterprise concerned.

The Type 1, CIM Architecture, on the other hand comprises only those aspects involved in the computer-based information technology of the enterprise, i.e. as noted here, those aspects included under the categories of Business Processing and Engineering and Control.

2.5 MAIN FEATURES OF ARCHITECTURES

As shown by Fig. 2.1 we have subdivided all enterprise activity into the following three areas.

- The Production Facility
- Human Organization Including
  Management
  Staff, and
  Production Functions
- Information Technology (Hardware and Software) including
  Business Information
  Engineering Systems
  Production Management, and
  Control Systems

What makes this subdivision useful is that it not only reflects distinct boundaries between the different types of enterprise activity but it also points out disciplinary (or cultural) boundaries. Using this division we can easily identify research and development results, vendors' product families, or standards which may be targeting to one or the other of the above areas.

For example, standards and product families have been developed for the integration of business information systems separately from the rest of the enterprise and are offered as solutions there because they have a generic structure which is applicable to virtually any case which could be considered. Similarly, large control system vendors have standards and product families which cover process control, product design, and production management, etc. The pattern is the same: these standards and families of products are based on some generic structure which is applicable to virtually any case of industrial control and engineering.

Although many of these generic structures are of use only in such well defined subdomains of all enterprise activities as those just noted they are nevertheless essential points of departure for the integration of the entire enterprise.

However, in order to realize the full potential of the use of each of these separate solutions one has to be able to form a unified structure that enables the integration of the information flow between these domains just as easily as within the confines of the domains themselves.

Type 1 Architectures (CIM Architectures) are for example a partial solution to such generic structures (see Fig. 2.1) since they allow the integration of the respective information technology regimes, such as business information processing system products with engineering (CAD, CAM or Control System) products thus describing a complete structure for the computer control system of the enterprise.

Given that either of these domains (business processing or engineering control (see Fig. 2.1) can separately define their own respective information processing, database, communications and human interface solutions (thus structures and standards), a CIM architecture which encompasses the control of all enterprise activities should have a common base rather then two separate solutions for the above mentioned subdomains. Type 1 architectures (or CIM architectures) as structural descriptions of the complete enterprise control system must therefore include traditional business processing as well as engineering, control and management in order to provide a proper solution to this requirement.

Most CIM Architectures known from the literature are of this Type 1 Classification and an overview of these is given in Chapter 5 of this book (e.g. [4,6] and many others, see references made there).

It is expected that the increasing drive to comply with open systems standards[1] and the adoption of client–server structures, will soon make it possible to establish enterprise-wide communication and processing infrastructures or Integration Platforms which will serve the needs of both the business processing and the engineering (engineering control and management) functions. Such integration platforms have already been developed for many of these limited domains. The remaining work is to extend them to complete the required overall generic platforms.

For example, IBM's System Application Architecture (SAA) [7,8,9] is such an intermediate platform devoted specifically to the domain of business processing. Any such integration platform, as a collection of hardware and software solutions, is preferably based on 'open systems' standards, allowing configuration flexibility and multivendor, vendor independent configuration of systems. These platforms must include a standard operation system interface [10], communication protocols [11,12], a user interface [13,14], a database access language [15], remote database access [16, 17], and distributed computing (client server computing) [18,19].

As an example of the overall capability desired, the Integrating Infrastructure of CIMOSA [3], which is under development, is intended to be a platform for the entire domain. Where practicable the AMICE Consortium developing CIMOSA has made a commitment to adopt existing standards such as those discussed above to promote these concepts.

---

1. Open systems standards include communications, operating systems, database access, distributed processing, and user interfaces.

2.6 TYPE 2 ARCHITECTURES IN RELATION TO TYPE 1

However, as is made abundantly clear by Fig. 2.1, the above discussion only comprises Item 3 of the list of three given in this section. A complete enterprise description by our definition must also comprise the integration of all human relations and human organizational activities along with those of the manufacturing facility, plus those of the computerized CIM system to have a completely integrated enterprise. Methods for carrying out the design work for each of the three domains are represented in Fig. 2.1 as horizontal patches adjoining the respective structural descriptions. Examples which may be given are: in the domain of business processing – A/D Cycle [20]; in the domain of production facility design – methods based on group technology [21,22,24] and on layout design techniques [23, 25]; in the domain of generic software engineering development – see [9, 26, 27]. There are also many different methods available for (separately) developing the human organization and management aspects of an enterprise.

An important message which will be derived from a reading of this book is that the simultaneous coverage of all of these areas is not yet available and that there are as yet only a very few (although quite important) attempts known in the open literature to try to develop such coverage. However, many groups worldwide are hard at work to correct this situation and important results can be expected soon.

We believe that the possibility to build comprehensive design methods in each area will accentuate the need for a complete life-cycle architecture to coordinate their comprehensive use together in a complete methodology. Naturally, such a development cannot be independent from the methods already developed for the separate domains as presented above. The substantial investment in these areas makes it necessary for the complete enterprise reference architectures to include, adopt and adapt these existing methods and tools as substantial building blocks in forming the overall complete methodology.

An overall and completely generic Enterprise Reference Architecture is thus an umbrella under which each of these separate disciplines and their methods can be integrated together with the others needed and one of the contributions which this book can make is to point out how such an architecture can be developed.

2.7 ORGANIZATION OF THIS BOOK

Chapters 3–5 present some background and supporting information concerning the integration architectures and modelling field necessary for our study. Chapter 3 defines the needs and requirements for architectures for integrating manufacturing activities and enterprises and thus establishes the goal against which our evaluations will be carried out. Chapter 4 presents a history of the field. Chapter 5 gives an extensive bibliography and discussion of the available architectures and models.

Chapters 6–8 describes the candidate architectures to be evaluated: CIMOSA (Chapter 6), GRAI-GIM (Chapter 7) and Purdue (Chapter 8).

Chapters 9–18 present the evaluation methods used and the direct results obtained. Chapter 9 introduces the background of the methods employed. These methods were: a detailed Questionnaire (Chapters 10 –13), its companion Short Form (Chapter 15), the answers to these as evaluated in Chapter 14; the Direct Mapping – One on One (Chapter 16); and the Matrix of Requirements Mapping (Chapter 17). An evaluation of the results obtained is presented as part of each chapter. Chapter 18 presents some additional observations concerning each of the architectures. Chapters 19 and 20 contain our Conclusions and Recommendations. First there is a compendium of the Conclusions derived from each of the evaluation methods and our analysis of these. Next we present our considered recommendations for the future of these architectures and for the development of the field itself. Also presented is a proposal for future work needed to further the field of enterprise integration and of architectures describing the field.

Appendix A contains the history of the Task Force and its work program, Appendix B lists its members in the first triennium of the Task Force's existence.

2.8 REFERENCES

[1] Doumeingts, G., Vallespir, B., Darracar, D. and Roboam, M., 'Design Methodology for Advanced Manufacturing Systems,' Computers in Industry, Vol. 9, No. 4, pp 271-296 (1987)

[2] Doumeingts, G., Vallespir, B., Zanittin, M., and Chen, D., GIM-GRAI Integrated Methodology, a Methodology for Designing CIM Systems, Version 1.0, LAP/GRAI, University Bordeaux 1, Bordeaux, France (May 1992)

[3] AMICE Consortium, CIMOSA Open Systems Architecture for CIM, Springer Verlag, Berlin (1993)

[4] Williams, T. J. (Editor), A Reference Model for Computer Integrated Manufacturing (CIM), ISA Publications, Research Triangle Park, NC (1989)

[5] Williams, T.J., The Purdue Enterprise Reference Architecture, ISA Publications, Research Triangle Park, NC (1992)

[6] Thacker, R. M., A New CIM Model – A Blueprint for the Computer-Integrated Manufacturing Enterprise, Society of Manufacturing Engineers, Dearborn, MI (1991)

[7] Anonymous, Special Issue on IBM's Systems Application Architecture (SAA) IBM Systems Journal, Vol. 27, No. 3 (1988)

[8] Radesi, S. J. and Czubek, D. H., SAA, IBM's Systems Application Architecture, Van Nostrand Reinhold, New York (1991)

[9] Thacker, R. M. and Dorfmann, M., 'System and Software Requirements Engineering', IEEE Computer Society Press Tutorial, p719 (1990)

[10] Lewine, D. A., Posix Programming Guide: Writing Portable Unix Programs With the Posix 1 Standard, O'Reilly & Assoc., Sebastopol, CA, pp 99 (1991)

[11]Anonymous, ISO International Standard DIS 7498 Data Processing – Open Systems Interconnection – Basic Reference Model, International Standards Organization (December 3, 1980)

[12] Anonymous, Manufacturing Automation Protocol, Version 3.0, General Motors Corporation, Society of Manufacturing Engineers, Detroit, MI (1987)

[13] Berlage, T., OSF/Motif: Concepts and Programming, Addison-Wesley, Reading, MA (1991)

[14] Davidson, A., Distributed Window Systems, A Practical Guide to X11 and Open Windows, Addison Wesley (1992)

[15]Anonymous, American National Standard for Information Systems – Database Language – SQL, ANSI Publication # X3.135-1986

[16] Anonymous, ISO Draft International Standard, Information Technology – Open Systems Interconnection – Remote Database Access, prepared by the ISO/IEC JTC1/SC21/WG3 (1992)

[17] Arnold, D., et al. 'SQL ACCESS: An Implementation of the ISO Remote Database Access Standard,' Computer (1991)

[18] Anonymous, Distributed Computing Environment (DCE), Open Software Foundation, Inc. (OSF) (1992)

[19] Anonymous, ISO International Standard 10746 Basic Reference Model of Open Distributed Processing (ODP)[1] Also as ITU-T (CCITT) Draft Recommendations for Open Distributed Processing X.900 (X.901, X.902, X.903 and X.904) In preparation: (July 1993)

[20]Anonymous, Special Issue on A/D Cycle, IBM Systems Journal, Vol 29, No. 2 (1990)

[21] Gallagher, C. C., Group Technology Production Methodology in Manufacture, Ellis Horwood, Chichester (1986)

[22] Ham, Inyong, Hitomi, Katsuo and Yoshina, Teruhiko, Group Technology: Application to Production Management, Kluwer-Nijhoff, Dodrecht (1985)

[23] Domschke, W. and Drexl, A., Location and Layout Planning, in International Bibliography, Springer Verlag, Berlin (1985)

[24] Snead, Ch. S., Group Technology: Foundation for Competitive Manufacturing, Van Nostrand Reinhold, New York (1989)

[25] Hales, H.L., Computer-aided Facility Planning, Marcel Dekker, New York (1984)

[26] Humphreys, W. S., 'The IBM Large Systems Software Development Process: Objectives and Directions', IBM Systems Journal, Vol. 24, No. 2, pp 76-78 (1985)

[27] Radice, R. A., Roth, N. K., O'Hara, A. L., Jr. and Ciarfella, W. A., 'A Programming Process Architecture', IBM Systems Journal, Vol. 24, No. 2, pp 79-90 (1985)

2.9 FURTHER READING

Anonymous, ISO International Standard Information Technology Software Life-Cycle Process, In preparation: IEC/ISO (JTC1) - SC7[2]

Date, C. J., A Guide to DB2, Addison Wesley, Reading, MA (1988)

Hoffnagel, G. F. and Beregi, W. 'Automating the Software Development Process', IBM Systems Journal, Vol. 24, No. 2, pp 102-120 (1985)

Martin, J. K., Chapman, Kavanaugh and Legen, J., SAA: Common Communications Support, Network Infrastructure Prentice Hall, Englewood Cliffs, NJ (1992)

1. As of August 1993 the ODP standard is in preparation. This four-part standard will consist of: Basic Reference Model of Open Distributed Processing: Part 1 Overview and Guide to Use, – Part 2: Descriptive Model, – Part 3: Prescriptive Model, – Part 4: Architectural Semantics.
2. As of August 1993

3

The needs of the field of integration

T.J. Williams

3.1 THE NEEDS OF THE FIELD OF INTEGRATION OF MANUFACTURING ACTIVITIES AND ENTERPRISES

The needs of world-wide industry today requires modification of manufacturing operations to assure: (1) a better and faster response to customers' requirements; (2) ever higher quality for products; and (3) increased flexibility and faster response in the introduction of new products and in responding to the needs of the marketplace. At this same time industry faces further requirements to increase their overall company earnings while: (1) decreasing the environmental impact of their factory's operations; (2) increasing the plant personnel and environmental safety of their operations; and (3) improving plant personnel working conditions and job satisfaction.

Integrated Manufacturing carried out with the aid of computers has been seen by many as the means by which much of the above could be accomplished. It is unfortunate that integration has generally (to date) been presented with a strong technology view only. For this reason efforts in this direction have only been partially successful, if at all, in achieving the management goals as expressed above. Often, these efforts have been too small and too narrowly focused and have resulted in nonintegratable "islands of automation" rather than the integrated whole originally conceived and desired.

At the same time this narrow focus has not been related to the business critical success factors of the company and thus has not made their proper impact on the potential benefits. Likewise the necessary human and organizational changes and the process technology issues have not been coordinated with, and fully integrated into, the information integration process itself.

There have been several major reasons for this. Primarily this has been due to the fact that those planning these projects have not realized the breadth and magnitude of the overall effort necessary and the resulting capital and other resources required. They have also not developed a total plan for the overall project necessary prior to commencing implementation, and thus have neglected to outline the total effort needed.

What is needed, therefore, for each company contemplating a major computer based integration effort is for the company to develop a Master Plan covering all of the anticipated effort required to integrate the whole of the company or factory operation.

After this, smaller projects within the monetary and personnel resources capability of the company can be initiated with the knowledge that the sum of this and all succeeding projects will result in the final total integration of the company's activities. This will be possible provided that the requirements of the initial planning effort or Master Plan be followed in each and every one of the resulting projects.

But the detail and effort required for even the Master Planning activity is itself large and if done improperly will only lead to difficulties later. Thus there is a need for a methodology to assure that the Master Plan as developed by a company's Integration Planning Team is complete, accurate, properly oriented to future business developments and carried out with the minimum of resources (personnel and capital) necessary. This methodology presents a detailed description of the tasks involved in developing the Master Plan including its continual renewal. It gives the detail necessary both to specifics and to quantity of information and data. It specifies the interrelationship of the informational, the human organizational and the physical manufacturing aspects of the integration considered; the management considerations and concerns; and the economic, cultural and technological factors involved as well as the details of the computer system required itself.

An Enterprise Reference Architecture models the whole life history of an enterprise integration project from its initial concept in the eyes of the entrepreneurs who initially developed it, through its definition, functional design or specification, detailed design, physical implementation or construction, and finally operation to obsolescence. The architecture becomes a relatively simple framework upon which all of the functions and activities involved in the aforementioned phases of the life of the enterprise integration project can be mapped. It also permits the tools used by the investigators or practitioners at each phase to be indicated.

Section 3.3 presents an extensive list of requirements and needs which the architecture and its associated tools should be able to fulfill during the development of the enterprise or manufacturing integration system including a list of questions which its use should be able to answer. In summary, the paradigm emphasizes the concepts of standardization, wide distribution, minimum duplication and ease of access and use (transparency) for the user.

Table 3.1 Definition of the new information paradigm

#	*Components of paradigm*
1	Object Oriented Methods should be used as much as possible
2	Standardized methods of database organization, management and system-wide access should be emphasized
3	Likewise standardized methods of communications should be used between sources, storage points and users of information
4	Shared data, data used by multiple functions or processes, should have a standardized, neutral format; anyone using that data should read from and write to that format and parse into any required specialized format for local use only as necessary
5	The methods of processing information requests and transmitting the results back to the user should be transparent to that user

Included herein, but separated out for emphasis in Table 3.1, is the new Information Paradigm now being widely accepted. This paradigm has been emphasized by CIMOSA in the principles espoused there. The architecture should, in short, be able to map concepts, answer and/or illustrate questions, as well as provide a means of comparing disparate systems, alternate implementation schemes for the same systems, and rival development methods and techniques [1–6]. Table 3.2 presents another version of some of the same requirements as a set of goals for enterprise integration architectures as derived from CIMOSA sources [2,3]. Table 3.3 continues this discussion of capabilities by listing a group of desirable modelling concepts which should be available with the architecture [2,3].

Table 3.2 Goals of a CIM architecture

Group[a]	*Goals*
1	Timely availability of the right information at the right place
2	Adaptability to the continuous change of the environment and the production processes
3	Flexibility of all enterprise processes and organizational structures
4	Processable descriptions of function and behavior of all activities within the enterprise for simulation and control purposes
5	Real time control of all enterprise processes
6	Most economic use of information technology
Group[b]	*Goals*
7	Possibility to make use of programs and machines from different vendors
8	Possibility to describe the CIM architecture by formal methods (rigorously defined syntax and semantics), to have a unique interpretation of the architecture which will allow one to perform any translation or perform any service desired

a. Realistically implementable in the next five years
b. Goals for the further future

3.2 NEEDS OF THE FIELD OF ENTERPRISE INTEGRATION BEYOND THOSE OF AN ARCHITECTURE ALONE

While the Task Force felt that the acceptance and/or further development of suitable architecture(s) was vitally important for the continuing growth and maturity of the field of enterprise integration, and its acceptance as the mode of optimization of the production and operation of manufacturing plants and other enterprises in the future, it recognized that architecture(s) alone will not accomplish this task.

Table 3.3 Modelling concepts important for use with enterprise reference architecture development (the following should be possible)

1	Verification of completeness and consistency for all described functions and objects (business processes, data, materials and resources including tools and fixtures) at any detailing level
2	Simulation of the enterprise model at any detailing level
3	Easy and fast change of the model in case of changing business processes, methods or tools
4	The use of the model to initiate, monitor and control the execution of the enterprise's daily operation
5	Repeated resource allocation during the execution of business processes to enable better and more flexible load distribution on the enterprise's resources
6	Model generation for existing enterprises as well as for enterprises to be built

Those factors listed below must also be brought about if the goals listed just above are to be accomplished. The architecture(s) will be extremely helpful in guiding the accomplishments of these factors, however. Enterprise integration, as a field, is itself an enterprise and the architecture(s) can be used to help plan and implement the growth and development of this enterprise as well as any other.

- Methods must be developed to make the use of the modelling techniques, tools, architectures, project development frameworks, etc, noted earlier as transparent as possible to users from a vast spread of different backgrounds and training. It must not be required that the ultimate users need to develop a computer science background to understand and apply them, ie, hide the theoretical complexity from the applications practitioner.
- Massive training programs must be initiated in unison with the transparency effort just noted so that the ultimate users can accept enterprise integration technology as a natural and easily assimilated part of the necessary background in their own application field, whatever that field may be.

- The further we can go in achieving Item 1 above, the easier will be the achievement of Item 2, and vice versa.
- One of the major difficulties in developing and implementing an enterprise integration program is the tremendous amount of detail required to completely specify, design and construct such a system. Means must be found to reduce greatly at least the "apparent degree " of the amount of detail required. Possible methods involve such techniques as configuration and parameterization rather than redesign or reprogramming, massive reuse of generic modules from past systems, etc.
- The ultimate goal must remain the achievement of a universal acceptance of the necessity of enterprise integration and its widespread accomplishment in all fields of endeavor, not just in industry, and especially not just in discrete manufacturing.

3.3 WHY DO WE NEED AN ENTERPRISE REFERENCE ARCHITECTURE?

To be valuable an Enterprise Reference Architecture should be able to supply answers or solutions to all or most of the following questions, requirements or situations:

1. The Architecture should present a method for the breakdown of all system functions to their inherent generic functions and tasks. These should be accomplished with a minimum number of types of basic building blocks. Conversely it should be possible to build up the overall functions from these ultimate basic building blocks.
2. The method should be able to show all existing relationships of those entities represented by the blocks or modules used for modelling.
3. The place of the human being in the enterprise system should be able to be explicitly shown, functionally and organizationally.
4. The Architecture should explicitly show the evolution of a CIM program or other enterprise undertaking from initial concept to and including operation as a completed system.
5. It should model the interfaces between humans and information systems, humans and manufacturing systems and between information and manufacturing systems as well as between foreign elements and other members within any system.
6. The Architecture should be able to model the effect of adaptation or change in the system in terms of limitations, effects, etc.

7. It should be able to show the limitations imposed by specific designs on the generic requirements for the system.
8. It should provide a basis for the development of needs for standards for various aspects of a system and be able to indicate their relative value.
9. The Architecture should be able to be used to map details of an enterprise or business entity (or a proposal or plan for same) against a similar mapping of an existing accepted standard to point out deficiencies, etc. in the existing or proposed system.
10. Similarly the Architecture should be able to map proposed future technology against present day best practice to show the suitability or desirability of the proposed future technology.
11. The Architecture should promote the concepts of flexibility, modularity and adaptability in the design, implementation and operation of the subject system being developed.
12. Wherever possible the Architecture should promote the concepts of the reuse of previous designs and programs in new systems.
13. Every effort should be made to develop and apply formal methods (such as rigorously defined syntax and semantics, for example) to describe and interpret the architecture and the representation of specific systems by it.
14. It should be possible to include economic questions in the subjects treated by the Architecture.
15. Most important, an Architecture should provide a common basis for discussion of a manufacturing integration or other enterprise system and thus assure a full and accurate interpretation by all discussing parties of all points under discussion.
16. The Architecture needs to reflect clearly the established enterprise decision making, its organization, its activities, its business processes, its information exchange and its material flows, in a form suitable for translation into an actual and processable implementation.
17. The Architecture must be amenable to a flexible modification, so as to reflect the changing enterprise environment, constraints and operations, and in particular to be capable of reflecting system implementations which are built in an evolutionary manner.
18. The Architecture and its supporting guidelines have to guide the user in the design, implementation and operation phases of such systems in a wide range of manufacturing industries.
19. The proposed architecture should be independent of existing technologies in manufacturing automation and computer science.
20. The proposed architecture should be independent of any given predetermined realization in terms of system configurations or implementations.

21. It is envisioned that several different views are necessary to completely describe a CIM system.
22. The overall system should be structured to limit the complexity of each entity. The resulting simplicity facilitates
    (a) Human comprehension
    (b) Computational load
    (c) Physical structure.
23. Shared data, data used by multiple functions or processes, should have a standardized, neutral format; anyone using that data should read from and write to that format, and parse into required formats for local use only as necessary.
24. All resources used in a manufacturing plant should be assigned a standard identification or classification code.
25. There is a need for a standard definition of terms related to data management, storage, communications, etc.
26. No distinction should be made between data and information in terms of the way in which it is transmitted and stored. There is obviously a semantic and cultural context to information in how it is used. Data shows no such context.
27. A heterogenous, multi-vendor environment of hardware and software with different levels of database management capabilities should be supplied.
28. Data is not necessarily tied to any particular control level; it resides in a database and is managed by a separate data administration system.
29. The hardware, software and database functions should be distributed throughout the CIM system as needed to maximize system speed of response capabilities. The system should thus follow the Principle of Locality leading to distributed processing and time-phased decomposition. Locality means that units in the same geographical region tend to work closely together. A unit generally works more closely with its neighbors than with those more distant.
30. For each data item in the system, there is some single entity that is responsible for that item. All data access between entities must be explicit. Data integrity requires explicit access both for reads and writes. Explicit access also controls system complexity and makes systems easier to maintain.
31. A system wide access method must be defined. The access method will support the controlled interchange of data between entities consistent with Item 30 above.

32. The architecture must provide the means for implementing the concept of data ownership as prescribed in the policies of the enterprise using the architecture.

(Items 23 and 32 relate to the need for a global database capability).

33. Human resources also need to be identified appropriately acknowledging the specific value of their functioning within the architecture.
34. The methods of processing information requests and then transmitting the results back to a user should be transparent to the users.
35. All users should have the same method of accessing information, regardless of where they are connected to the network.
36. Interfaces to the outside world (customers, suppliers, government agencies, capital providers, etc.) should be standardized.
37. Interfaces, protocols and interactions across all levels of management, support and operational systems should be standardized for control, data storage, retrieval, and communications.
38. All major functions should have a standard format for I/O which would be used by any vendor supplying software to perform that function.

(Items 34–38 relate to the OSI or Open System Interconnect feature now being widely requested of vendors).

39. The CIM system should be functionally organized to avoid the specification favoring any one physical embodiment.
40. The CIM system should incorporate sufficient fault tolerant, redundant, fail safe, or fail soft features to assure the safety of personnel, plant equipment and environment under any possible foreseen plant episode.
41. The appropriate relationship between modularity and system functional performance must be developed. It should assure the easy replacement of any modules with appropriately enhanced successors without jeopardizing the interconnectivity of the changed functional capability with other neighboring functions. The resulting system should uphold the Principle of Autonomy, i.e. leading to a modular system with high cohesion and low coupling. Autonomy means that individual modules are as independent in action as overall integration can permit.
42. The specification for the architecture must be open ended in its ability to be extended and in its ability to encompass new technologies without unreasonably invalidating current realizations.
43. Recovery procedures must be standardized for dealing with the evolutionary nature of these systems (i.e. additions and deletions to the management, support and operational system for both hardware and software).

44. The architecture should have a generic part which can be applied to a wide range of manufacturing operations and organizations.
45. Data passed between entities is of three types:
    (a) Raw data, which emerges directly from a sensor.
    (b) Clean data, which has been smoothed and conditioned.
    (c) Information which has been associated with a semantic context – as noted in Fig. 3.1.

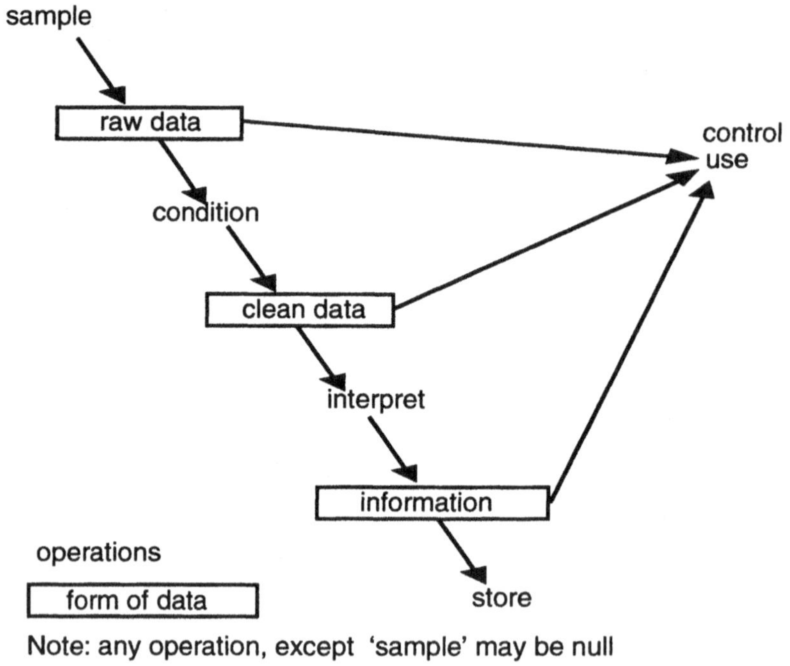

Figure 3.1 The different stages of data reduction to information and storage.

3.4 REFERENCES

[1] Doumeingts, G., Vallespir, B., Darracar, D. and Roboam, M., 'Design Methodology for Advanced Manufacturing Systems,' Computers in Industry, Vol. 9, No. 4, pp 271–296 (December, 1987).
[2] ESPRIT Project 688 AMICE, Open System Architecture for CIM, Springer Verlag, Berlin (1988).

[3] ESPRIT Consortium AMICE, Open System Architecture, CIMOSA, AD 1.0, Architecture Description, ESPRIT Consortium AMICE, Brussels, Belgium (1991).
[4] Industry-University Consortium, Purdue Laboratory for Applied Industrial Control, An Implementation Procedures Manual for Developing Master Plans for Computer Integrated Manufacturing, Report Number 155, Purdue Laboratory for Applied Industrial Control, Purdue University, West Lafayette, IN (June 1992).
[5] Project Staff, Tasks and Functional Specifications of the Bhilai Steel Plant Integrated Control System (INCOS), Volume I – Chapters 1-12, Volume II – Chapters 13–25, Volume III – Chapters 26–31, Purdue Laboratory for Applied Industrial Control, Purdue University, West Lafayette, IN (April 1986).
[6] Williams, T. J., Editor, A Reference Model for Computer Integrated Manufacturing, A Description From the Viewpoint of Industrial Automation, CIM Reference Model Committee, International Purdue Workshop on Industrial Computer Systems, Purdue University, West Lafayette, IN (1988); Instrument Society of America, Research Triangle Park, NC (1989).

4

State-of-the-art on models, architectures and methodologies

G. Doumeingts and D. Chen

This chapter intends to give an overview on models, architectures and methods for the CIM systems design. We start from presenting the basic concepts and definitions of CIM, model, architecture and methodology, as well as the concepts of the economic and human aspects of CIM. Then we present some early work performed in the 1970–80s; ICAM, CAM-I, NBS, SADT, SSAD, GRAI, MERISE and SSADM. Following this there is a state-of-the-art evaluation of the work performed within the ESPRIT Program, particularly MMCS, E932 knowledge based factory model, PAC model, CIMOSA, IMPACS and FOF. We also give an overview on recent CIM architecture work in the USA, and in particular the Purdue Reference Model and Enterprise Reference Architecture and the EIP approach. Some IT vendor based approaches are also presented such as the IBM model and the Digital model. The CIM architecture work performed in the Asia and Pacific area is not ignored. Simulation models for manufacturing systems are also mentioned. The object oriented approach for CIM systems design performed by IPK Berlin, Germany and recent developments on the GRAI-GIM methodology and ECOGRAI at the GRAI laboratory in France are also reported.

4.1 BASIC CONCEPTS AND DEFINITIONS

The term CIM is usually used to describe the process of integration of all of the elements involved in manufacturing, by means of computer techniques, and has usually been confined to a purely technical interpretation. Today, the many difficulties encountered in the course of various implementations have shown that such a narrow outlook is not satisfactory: economic, social and human aspects must be taken into account. We consider that 'CIM refers to a global approach – in an industrial environment – which aims at improving industrial performance. This approach is applied in an integrated way to all activities, from designing to delivery and after sale, and uses various methods, means and techniques (computer and automatic techniques) in order to simultaneously improve productivity, reduce costs, meet due dates, increase product quality, secure flexibility at the local or global level in a manufacturing system, and involves every actor. In such an approach, economic, social and human aspects are at least as important as technical aspects' [1]. This broad definition shows that CIM cannot be bought; each firm must devise its own, which explains why methodologies must be made available so that CIM systems can be built.

The word 'methodology' means a set of methods which involve the use of: modelling formalisms and their associated graphic tools; models and reference architectures; and structured approaches.

A modelling formalism is a means to represent pieces of knowledge that have to be transmitted unambiguously. It allows one to build models according to a set of associated concepts. The theoretical basis for modelling formalisms can be found in the graph theory, the languages and logical structures theory, etc. The modelling formalisms that are used to model production systems are often associated with graphic tools. A good diagram is often better than a long speech.

A model is an abstract, simplified representation of reality. A model of a CIM system can only represent a set of selected elements concerning the domain studied, and in agreement with defined objectives. A good model amplifies the important characteristics and conceals the details which are considered to be of a low or no importance at a given abstraction level. In the CIM area, models are supported by mathematical formalisms, languages and/or graphical tools.

A CIM architecture can be defined as a structured set of 'models' which represent the invariant building blocks of the whole CIM system. It is considered as a basis for the design and the implementation of CIM systems. These models, which contain the invariant elements and the relations between these elements, describe respectively the 'what' (what a CIM system is conceptually composed of), and the 'how' (how a system technically works), and show how to transform the models into realities, i.e. the working system [1,2]. A reference architecture is the architecture for a class of manufacturing systems. One derives the architecture of the enterprise studied by adapting the reference architecture to the specifics of the enterprise.

Generally speaking, a structured approach defines a set of steps to be followed when applying a methodology to the solving of a problem. In a CIM design methodology, the structured approach should cover the whole life-cycle of the CIM project which is split up into several steps (analysis, design, development, implementation, operation). All these steps should be perfectly defined and based on the project type structure, consisting of particular actors whose roles are clearly defined. For each of these steps, models will be built and coherence and validation will be controlled between the various steps, and between the various models.

Talking about the economic issue of CIM one can find that in today's companies the economic factor is often not taken sufficiently into account in everyday decision making. Many of the management decisions are made without knowing their effects on costs. Recently, the GRAI laboratory has developed the ECOGRAI method which allows one to identify the economic decision centers of a production management system in order to implement performance measurement systems.

Implementing a CIM system in a company will inevitably cause large changes in its running style. During the ESPRIT Project 1217, the concept of HC-CIM (Human Centered CIM) was developed. Traditional functional system design considers the human as a mechanism basically equivalent to other resources like machines, etc. The HC-CIM approach provides an opposite understanding of the human role in the production process; not as a supplement to supervise the process and manage exception, but as the acting and interacting backbone supplemented and supported by technologies to make the human capabilities more productive. It requires the integration of the technical and the social domain and rejects a technology driven understanding of integration [3].

4.2 EARLY WORKS ON MODELS OF ADVANCED MANUFACTURING SYSTEMS

Between the end of the 1970s and middle of the 1980s, some work on models of advanced manufacturing systems was performed in the USA and in Europe. The ICAM model (Integrated Computer Aided Manufacturing) was elaborated within one of the major projects led by the US Air Force. The ICAM model [4], also called 'Model of Manufacturing' was especially designed for the aerospace industry. It displayed the functions of aerospace batch manufacturing and the detailed relationships between those functions. The ICAM model was supported by The IDEF (ICAM Definition) methodology. A more general model for a large variety of manufacturing enterprises was built within CAM-I (Computer Aided Manufacturing International). CAM-I is a nonprofit organization formed to promote cooperative R&D efforts in CADCAM. Over 100 large companies from around the world have joined CAM-I. A factory model was proposed with four levels of decomposition [5]. At each level we find seven main generic functions. Two levels of this model (shop and cell levels) were described in the 'Factory Management' project in which GRAI laboratory was involved. However, no associated tools were developed within CAM-I. The CAM-I model was found to be too conceptual and was not widely used.

Another initiative launched during the same period of time was the NBS model which later became a widely accepted reference model. It also the first model published as a CIM architecture. The NBS model aimed to have an extreme flexibility in order to emulate the wide variety of manufacturing cells which are typical of a small machine job shop in the domain of machining and assembly. It fits particularly well in a fully automated manufacturing environment with a highly complex computer based sensor system [6]. At the same period in Europe, the GRAI models [7] were developed to meet the needs of the modelling of decisional systems. The GRAI models consist of a macro conceptual reference model for manufacturing systems (Fig. 4.1) and a micro conceptual model for decision centers. Compared with the ICAM, CAM-I and NBS models, GRAI models are based on system theory and present the decision making aspect. The GRAI model is a reference through which various elements of real world can be identified. The macro conceptual model is used to express one's perception and ideas on the manufacturing system which is decomposed into a decision subsystem, an information subsystem and a physical subsystem. Particularly within the decision subsystem one finds a hierarchical decision structure composed of decision centers. Decision centers are connected by a decision frame (objectives, variables, constraints and criteria for decision making). The operating system is an interface between the decision system and the physical system. The micro conceptual model is used to represent the internal elements and structure of the decision center.

4.3 METHODS FOR DESIGNING ADVANCED MANUFACTURING SYSTEMS

In the middle of the 1970s, there was a need to deal with complex systems and structured analysis and design techniques began to draw attention. SADT (Structured Analysis and Design Technique) was developed by SOFTECH to analyze such complex systems [8]. SADT uses the Actigram (similar to IDEF0) and the Datagram for studying function and data aspects.

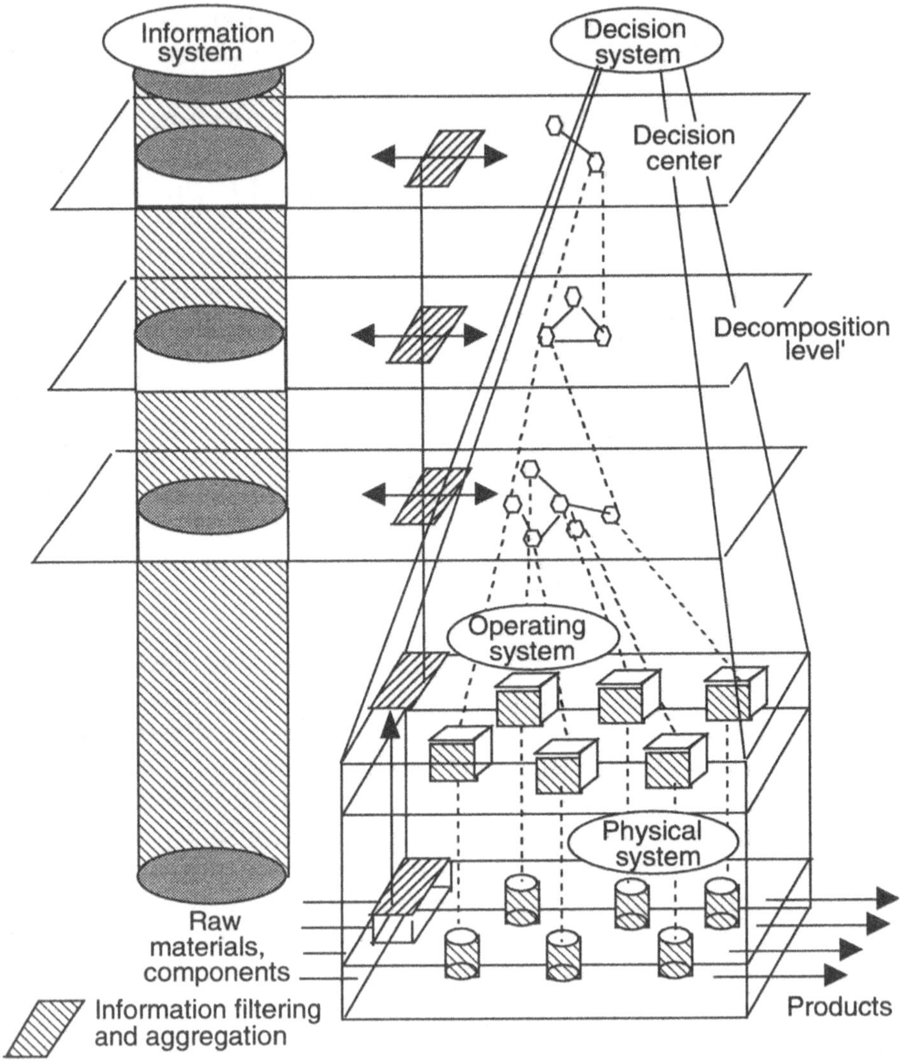

Figure 4.1 The GRAI conceptual model.

The method is supported by many computerized tools. SADT is not, properly speaking, a method to design manufacturing systems. However, it has the advantage of proposing a relatively general approach which can be adapted to a great number of applications. SADT turned out to be more adapted to the analysis and understanding of a complex system than to the technical designing of a manufacturing system. At the same period, the

SSAD (Structured System Analysis and Design) method was developed by Chris Gane and Trish Sarson [9] and marketed by Improved System Technologies Inc. (I.S.T.), now owned by Mac Donnell Automation. It is derived from the works of De Marco on the Data Flow Diagram [10]. The method consists of building a graphical (non physical) model of systems. Compared with SADT, SSAD is particularly well adapted to the analysis and design of computer supported information systems. However, SSAD cannot be considered as a method for designing CIM systems.

In order to cover various other aspects involved in the design of manufacturing systems, the ICAM project developed the IDEF (ICAM Definition) methodology to support the elaboration of ICAM models. IDEF in total consists of several different modelling approaches: IDEF0 is used to build a functional model, IDEF1 is used for the information model and IDEF2 for a dynamic model. Recently, IDEF1 was extended to become IDEF1x to help build a semantic data model. IDEF3 and IDEF4 were also developed to help deal with the topics of the enterprise scenario and object oriented modelling. In IDEF, structured approaches with well identified team works are defined. IDEF is the most used approach for CIMS design. IDEF0 is closely related to SADT.

Similar work on methods was also carried out in Europe, notably the GRAI method [7] and MERISE method [11]. Before developing the GRAI method, some existing works had been reviewed, notably SADT and SSAD methods. It was found that the decisional aspects were not very well taken into account in these methods. So, it was important for the GRAI method particularly to deal with the decisional aspects of manufacturing systems. Based on the GRAI models, two formalisms were developed to model the macro decision structure and the micro decision center; the GRAI grid and GRAI nets. A structured approach was defined to show how to apply the method. The GRAI method has been transferred to industry (over 100 industrial applications). Two ESPRIT-1 Projects (418 and 932) allowed some improvements of the method, and several ESPRIT-2 Projects are using it, such as ESPRIT 2338 and 2434, and an ESPRIT basic research Project FOF [18]. While the GRAI method focused its attention on the decisional aspect, the MERISE method was developed in France for the analysis and design of information systems. It is used for designing data models (static models) and data processing models (dynamic models) at three levels of abstraction; conceptual, logical, and physical.

A formalism, based on the principle of Petri nets, has been developed to build data processing models. The Entity/Relationship formalism has been chosen to model static data structures. The originality of the MERISE method probably lies in the fact that it considers the synchronization of data in the data processing model as a vital aspect in designing an information system, whereas quite a number of other methods (SADT, SSAD) do not regard the synchronizing aspect as a fundamental factor. In MERISE, abstraction levels are clearly defined, while in other methods, they do not exist.

Another method dealing with information systems design is the SSADM (Structured Systems Analysis And Design Method) developed in the UK by CCTA (Centre Computer and Telecommunications Agency) in the early 1980s. It is the UK government's standard method for carrying out the systems analysis and design stages of an information technology project. SSADM has been traditionally used for the development of medium or large systems. However, one variant of SSADM is 'Micro SSADM' which is for small systems. SSADM starts from defining the information system strategy and then develops a feasibility study module. These are followed by requirements analysis, requirements specification, logical system specification and a final physical system design.

4.4 RECENT WORKS ON MODEL/ARCHITECTURES WITHIN THE EUROPEAN ESPRIT PROGRAM

In the middle of the 1980s, the Commission of the European Community launched the ESPRIT research program. The GRAI Laboratory was involved in several ESPRIT Projects in the CIM area, particularly ESPRIT Project 418, Open CAM System; ESPRIT Project 2338, IMPACS; ESPRIT Project 3143, FOF; and ESPRIT Project 5288, CIMOSA.

The MMCS (Manufacturing Management Control System) [13] was developed within the ESPRIT Project 418 – Open CAM System, in which the GRAI laboratory was involved. This model is aimed at the problem of Shop and Cell planning and control with well defined interfaces (see Fig. 4.2). These interfaces transform, where necessary, the external data view into the MMCS view. Via a database, the system is able to interface with the available commercial MRP system at the factory level. The core functions of the MMCS are the Shop and Cell Controllers. Both the controllers have the same structure. They are responsible for the management of jobs, capacities and materials at Shop and Cell Levels respectively. While the MMCS model is very generic it is challenged by an even more generic model: the PAC (Production Activity Control) model. The PAC model was developed

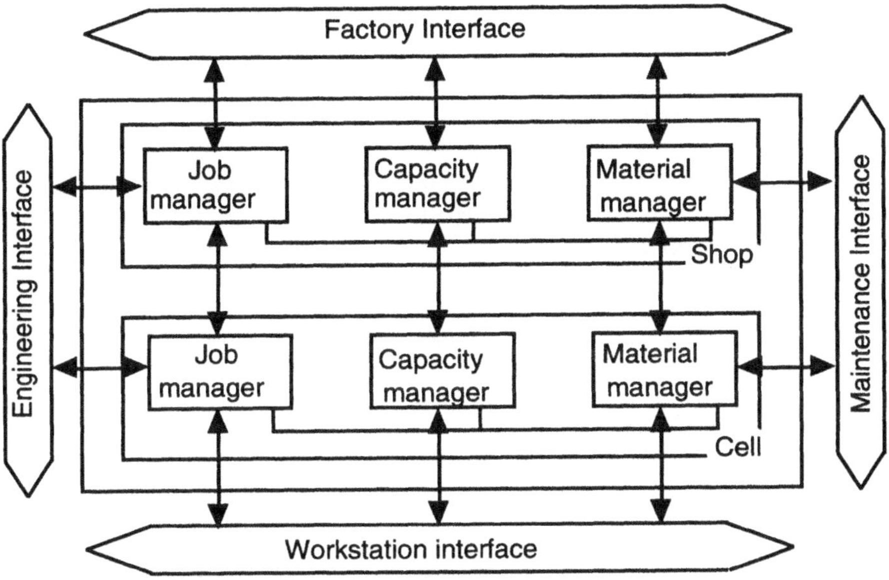

Figure 4.2 The MMCS model.

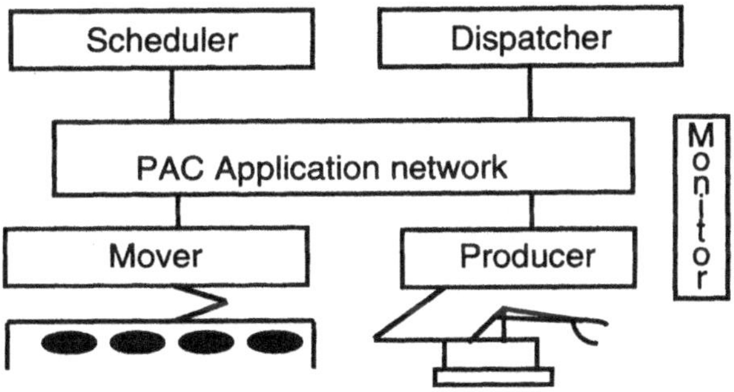

Figure 4.3 The PAC model.

within the ESPRIT Project 477 entitled COSIMA (Control System for Integrated Manufacturing) [14].

One of the basic principles of PAC as shown in Fig. 4.3 is the recognition of five basic building blocks representing the functions required for production control: Scheduler, Dispatcher, Mover, Producer and Monitor. Communication between these building blocks is facilitated through the use of the Application Network which was developed as part of the project. This

Application Network also supports message passing across different nodes thus providing the means to develop a distributed PAC system. MMCS and PAC models are limited to the shop and cell levels. A more global model for factory supervision and control was defined in the ESPRIT Project 932 to support decision making at all levels. This model is based on the NBS model and the GRAI model and uses an intelligent workcell controller concept. A controller (see Fig. 4.4) performs three basic tasks (Planning, Quality Control and Preventive Maintenance). Various expert system tools have been developed to support these controllers.

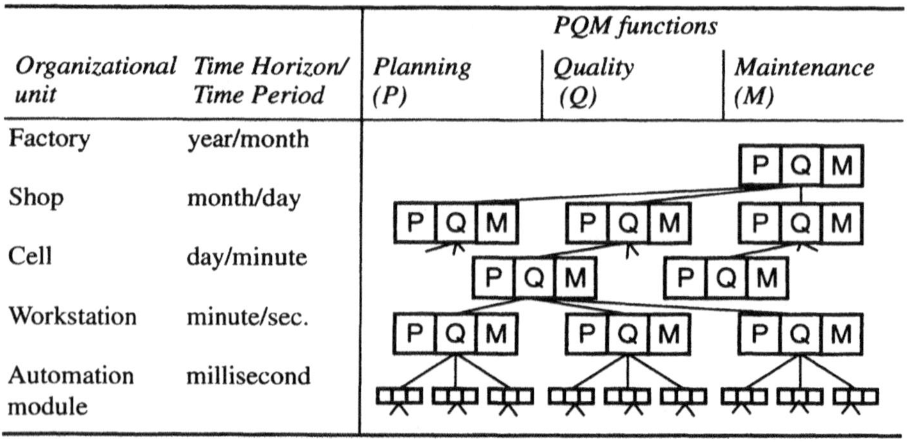

Figure 4.4 Knowledge based real-time model.

The knowledge based, real-time control, reference model enforces the decision making aspect by introducing some important concepts from the GRAI model, such as the Decision Horizon/Period and the Decision Frame, etc. The implemented system becomes totally goal oriented by means of the use of a 'decision frame'. The ESPRIT Project 932 finds its sequel in a new project; ESPRIT Project 2434 in ESPRIT 2.

The ESPRIT Project 2434 is concerned with real-time controllers for distributed factory supervision. The results achieved in ESPRIT 932 were used directly in ESPRIT 2434. The objective of this latter project is to make modern production philosophies (OPT, JIT, LOP, etc) available on the factory floor by delivering decision support to each relevant function of the factory using knowledge based software techniques.

During the same period, another initiative for developing an Integrated Manufacturing Planning And Control System was launched within the frame of the ESPRIT Project 2338 entitled IMPACS. It is probably the first attempt to design integrated planning tools that would bridge the gap between global planning and production control based on very long term strategic planning (3 to 5 years) and real-time control at the cell level [15].

IMPACS considers it very important to design all of the tools needed within an architecture context. A framework for an architecture [16] (see Fig. 4.5) and the IMPACS method also known as GRAI-GIM (General IMPACS Method) have both been proposed. The IMPACS architecture is

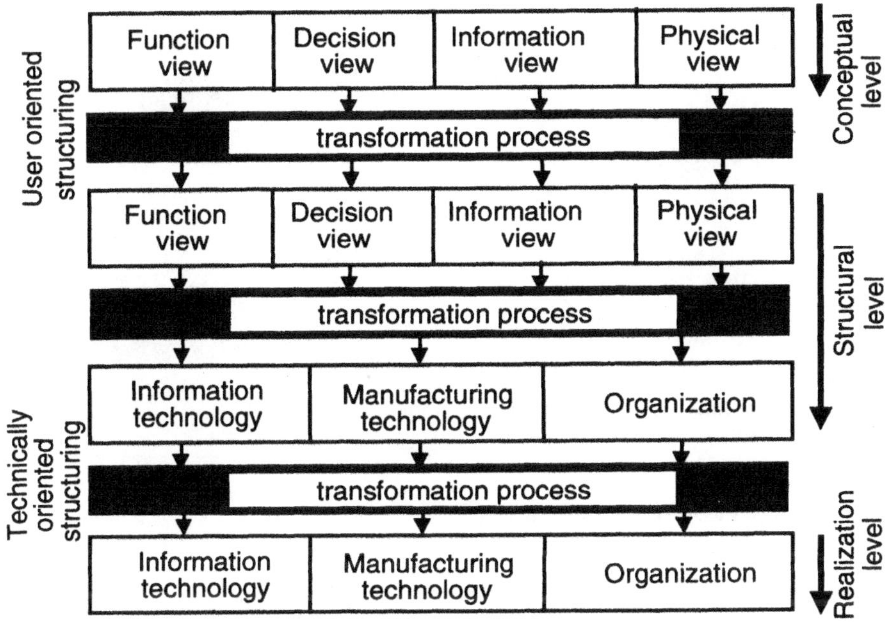

Figure 4.5 Framework for the IMPACS architecture.

defined at three levels of abstraction, conceptual, structural and realisational. The conceptual architecture allows one to define all the elements that are needed in an IMPACS system; the structural architecture defines how to structure the identified elements into the conceptual architecture, and the realisational architecture shows the final real tangible system.

It is generally considered that production in the future will be based on a one-of-a-kind system of product production. That's why the FOF Project (Factory of the Future – a Basic Research Action in the ESPRIT Programme) was launched. The ultimate goal of the project is to obtain a designer's workbench for the development of CIM in future production systems providing one-of-a-kind products (OKP) to the market, each OKP will require its own product design and process design [12].

There are many methods for describing the operations which occur in production systems. However, the methods are all fundamentally fragmented as they are all based on a set of scattered, theoretical notions (communication, organization, human point of view, decisional approach). Fig. 4.6 illustrates the various 'views' identified in the project, and points out

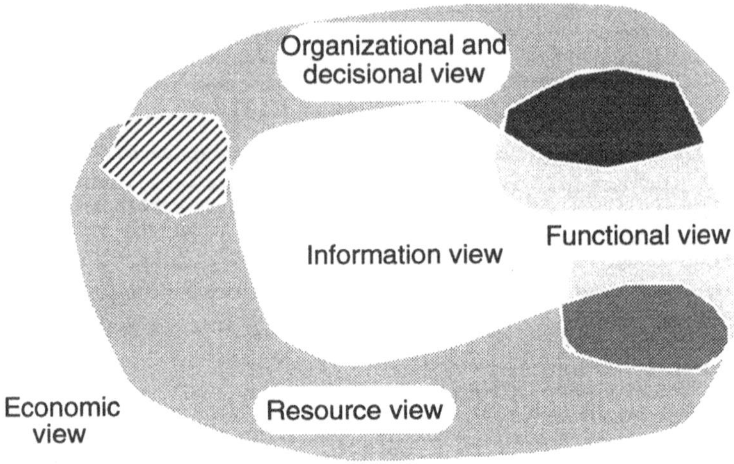

Figure 4.6 Views of the FOF model.

their relationships. As the views are presented separately, we need an integrated approach for PMS design. This integrated approach in the design phase was first based on an integrated model called REMBRAND. A set of Design Choice (DC) and a set of Performance Indicators (P.I.) were connected by a set of relations. The designs of the Production Management System (PMS) were done in agreement with the value of the P.I. and depending on the DC. See Fig. 4.7.

In some more specific areas of CIM, it is worth noting that the ESPRIT Project CNMA allows the extending of MAP. The ESPRIT Project CIM-ALIVE established a common project CIM reference model identifying cost effective reusable CIM concepts. ESPRIT Project CIM-Plato developed a prototype toolbox consisting of a set of computer-aided procedures and tools to design, plan and install FMS and FAS systems within the CIM environment.

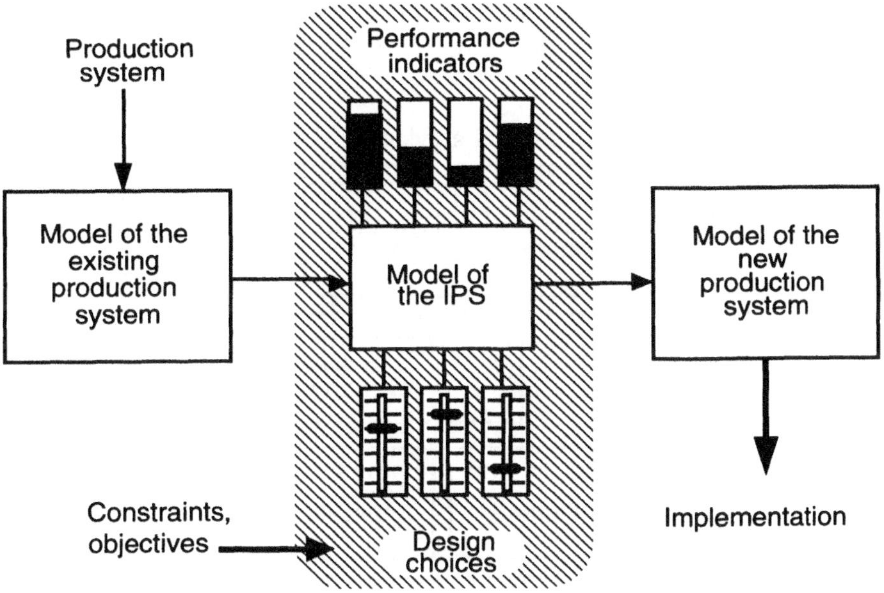

Figure 4.7 FOF workbench: relation between DC and PI.

In particular, in the methodology area, the ESPRIT Project 2706, MICIM (Methodology for the Introduction of CIM), considered that the methodology for progressing towards an integrated manufacturing environment has largely been neglected with most of the emphasis of current work being placed on the problems arising in technology. This project therefore concentrated on defining a methodology to remedy this. On the economic aspect, the ESPRIT Project CIMSIM intends to provide a methodology and associated tools for the economic and technical evaluation of various CIM solution options. An ESPRIT Exploratory Action 5603 is dealing with the problems of incorporating human factors into technology research and development along with techniques for the economic justification of CIM.

Lastly, we close this section by introducing CIMOSA [17], the well known CIM Open System Architecture developed by the AMICE Consortium, and the most important CIM initiative within the ESPRIT program. The term AMICE is a reversed acronym for 'European CIM Architecture'.

The aim of this project is to elaborate an open system architecture for CIM and to define a set of concepts and rules to facilitate the building of future CIM systems. An important aspect of the project is its direct involvement in standardization activities. The two main results of the project are the Modelling Framework, which is well known, and the Integrating InfraStructure (see Fig. 4.8). The Modelling Framework

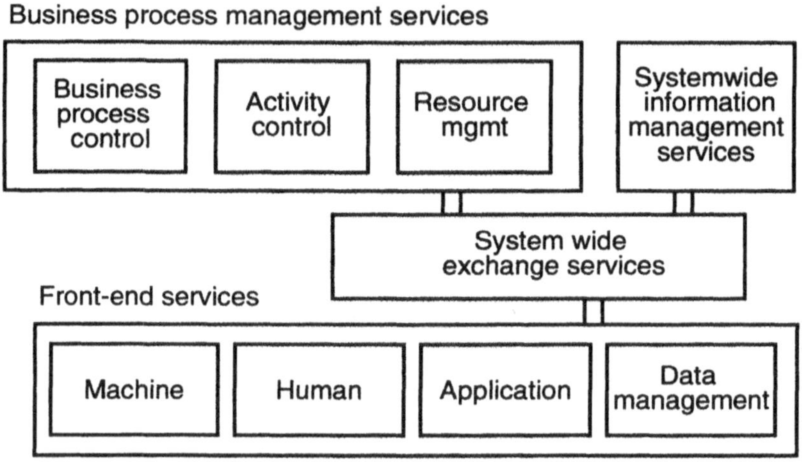

Figure 4.8 CIMOSA Integrating Infrastructure.

supports all phases of the CIM system life-cycle from requirements definition, through design specification, implementation description and execution of the daily enterprise operation. The Integrating Infrastructure provides specific information technology services for the execution of the Particular Implementation Model, but what is more important, it provides for vendor independence and portability.

CIMOSA insists particularly that the released model of the CIMOSA architecture should be processable or executable and evolutive, e.g., it can be modified easily during the run time. CIMOSA also intends to provide a methodology to show how to use the reference architecture to get a particular architecture of the studied enterprise. These are known respectively as the CIMOSA Model Creation Processes, and as the CIMOSA System Life Cycle. The CIMOSA modelling framework has been adopted by CEN/ CENELEC as the European prestandard for enterprise modelling, where it is known as ENV 40003.

It is worth noting that several ESPRIT projects are currently carried out to evaluate and validate CIMOSA, notably the ESPRIT Project VOICE (Validation OSA in Industrial CIM Environments), the ESPRIT Project CIMPRES (CIM Model and Implementation Concept in Precision and

Special Tooling Industry). Another ESPRIT Exploratory Action, 5605, is developing applications of CIM concepts, architectures and technology to Computer Integrated Agriculture based on CIMOSA.

4.5 RECENT CIM CONCEPTS AND DEVELOPMENTS IN THE USA

For several years, the distinct CIM architecture developments in the USA have been abundant, but each of them is based on its own architectural definition. No consensus has been achieved on what a CIM architecture is. There is no way of justifying an architecture proposal. The CIMCON 90 Conference at Gaithersburg showed that this situation had not been improved as of that date. Thus the USA parallels the situation in Europe.

CAM-I [5] considers the content of a CIM architecture was not intended to provide 'the answer' to the many complexities of CIM implementation, but rather was to be used to establish a common understanding of some of the identified elements comprising the CIM jigsaw puzzle. A CIM architecture should be viewed as a blueprint for the future, not as a detailed migration strategy or strategic planning road map to solve a current problem [18].

NIST considers it necessary to define a global architecture to integrate several separated architectures. They have argued that ‘such an architecture has three interrelated management components, production, information, and communication’ [6]. For each of these components, there must be an architecture. As for the Battelle Memorial Institute, they consider that there are many parts to defining an architecture for a CIM system. The following categories have been identified: functional relationships, information relationships, processing relationships, sequence relationships and control relationships [19]. The RAMP (Rapid Acquisition of Manufactured Parts) architecture which they developed is an aggregation of several perspectives of the RAMP system, including an IDEF0 functional model, a Yourdon-De Macro data model, a simulation model, a process control flow model, and physical models of the shop floor and hardware and software configuration [20].

From the point of view of Philips USA, a CIM architecture should describe a production organization as a configuration of interacting components. They assume that the purpose of an ‘architecture’ is to show how a production organization can be integrated, i.e. composed of closely coordinated components, and controlled. They indicate that a common, although not universal, denominator of CIM architectures is their hierarchical structure of ‘controllers’. They have proposed that a CIM architecture should have a ‘natural structure’, e.g. a structure with the ‘conceptual integrity that allows a human to master a production organization as a whole, despite its inherent complexity' [21].

The Sandia National Laboratories, a subsidiary of AT&T, considers that a CIM architecture is a formal representation which provides an integrated user interface over a very broad collection of application platforms, all from different vendors. There are several sources for obtaining physical and operational descriptions of computer and network architecture, but for a functional system architecture only the guidance exists and not the description available for the more specific system. They consider that there are only two useful viewpoints in the logical representation of a system architecture. These are the User View and the System View. The system view might be further represented by both a Designer's view and an Implementor's View [22]. They have discovered that one must not only base a new CIM architecture upon functional requirements and specifications, but that these must, in turn, rest upon well considered business and technical policies [22].

From Honeywell's point of view, 'the architecture does not provide a solution; it provides the discipline and the mechanisms needed to evaluate various alternatives, and to select a solution with reasonable confidence' [23]. The CIM reference architecture provides a common starting point for all the manufacturing units in an organization to use in developing their own organization specific architectures. They define the CIM architectures as a structured interrelationship of principles, functional models and guidelines.

The Purdue University group has proposed a 'Purdue CIM Reference Model' [24]. It was the oldest of the reference models but has recently been updated to meet the needs and requirements of CIM modelling specifically. They consider that a reference model is a previously agreed-upon or 'standard' definitive document or conceptual representation of a system. The reference model defines requirements common to all implementations, but is independent of the specific requirements of any particular implementation. In May 1988, the CIM Reference Model Committee of the International Purdue Workshop on Industrial Computer Systems completed this Reference Model for CIM: a Description From theViewpoint of Industrial Automation [25]. This reference model compiles a generic listing of those tasks which must be carried out by the integrated computer control system of a manufacturing plant in any industry to achieve CIM as understood today. The model arranges these tasks in a hierarchical functional framework to show their proper primacy and subordination to each other. It also shows the necessary data flow between these tasks, defines the relationship of the factory to those external entities which influence its operations, and develops a mechanism which allows the definition of the procedure for the carrying out of any one task [24].

The Purdue group has also very recently developed a major extension to their previous work to develop a full architecture called The Purdue Enterprise Reference Architecture. This has been published by the Instrument Society of America [25] and is becoming widely accepted in America particularly in process industries circles.

Another approach drawn to our attention is the CIM architecture studied in Carnegie Mellon University. It uses the intelligent networking concept as the element to facilitate the integration of the distributed heterogeneous functions of a manufacturing enterprise [26]. The integration is supported by having the network; first, play a more active role in the accessing and communication of information, and; second, by providing the appropriate protocols for the distribution, coordination, and negotiation of tasks and outcomes. The Enterprise Management Network architecture [27] uses the GRAI approach for its organization layer.

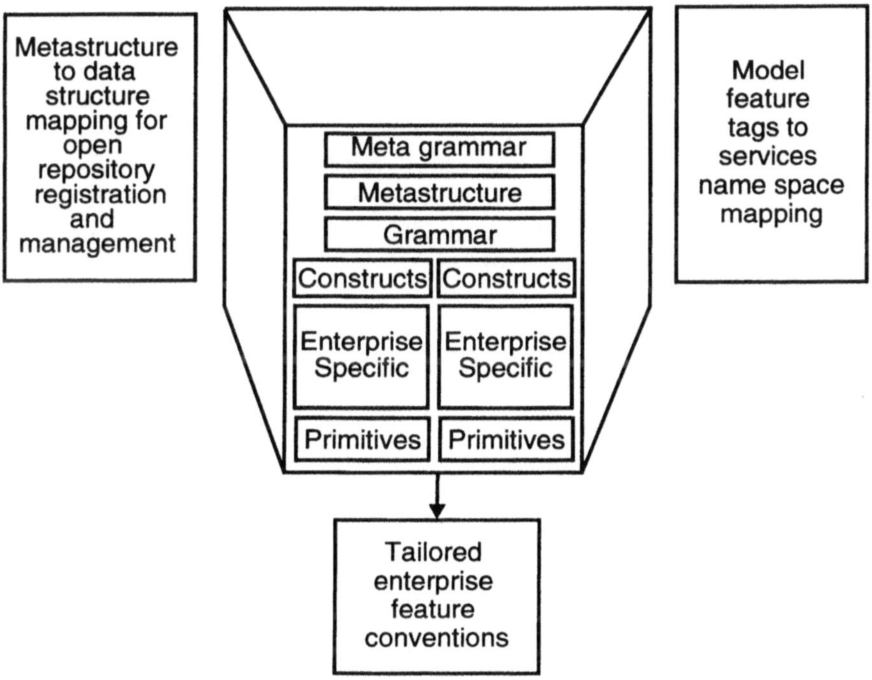

Figure 4.9 Unified model strategy.

Besides the university and company developed architectures, the EIP (Enterprise Integration Program), a research and development program composed of several projects, is being carried out in the USA. The Program Manager is the US Air Force and the Prime Contractor is SOFTECH. The EIP has four phases (framework, 18 months; commercialize, 24 months; pilot sites, 14 months; round up, 4 months). EIP considers that there is a global race for the technology to be first to market with the best. The range from the strongest to weakest will be decisive. The best performers will be those who excel at coordination. Specific improvements in process or product yield will only give temporary advantage. The source of a sustainable advantage will be the ability to continually improve.

EIP believes that Enterprise Integration is concerned with getting the right information, parts, processes, people, cash and products to the right places at the right times, and with reacting rapidly to changes in market demand, resources, technology and organization. The EIP seems to be the most important program in the US for the time being in the area of CIM. Close cooperation with the European AMICE Consortium (CIMOSA project) has been established in order to seek the complementary advantage which may exist between EIP and CIMOSA.

It seems that EIP addresses a broader scope than that of CIMOSA. EIP intends to integrate any and all models in an enterprise, all running in a heterogeneous and open environment (See Fig. 4.9). The US had sponsored two studies on an Enterprise Integration Framework (EIF) which focused on the applicability of CIMOSA to the US definition of the EI problem space. Cooperation between AMICE and US would emerge in a more internationally harmonized framework.

4.6 MODELS DEVELOPED BY IBM AND DIGITAL

The two biggest IT vendors in the United States, IBM and Digital, early realized an enormous market potential for CIM products and started to concentrate on the development of CIM solutions in the early 1970's. Presented below is only an overview of part of their known work.

In the early 1970's, IBM begun to develop a general CIM philosophy named Communication Oriented Production Information and Control System (COPICS) which comprised all the major planning and control activities of manufacturing. The emphasis was made in communication, database management and presentation as shown in Fig. 4.10. Originally IBM intended to promote their software and hardware solution within the framework of COPICS. No general CIM architecture was developed. Due to the

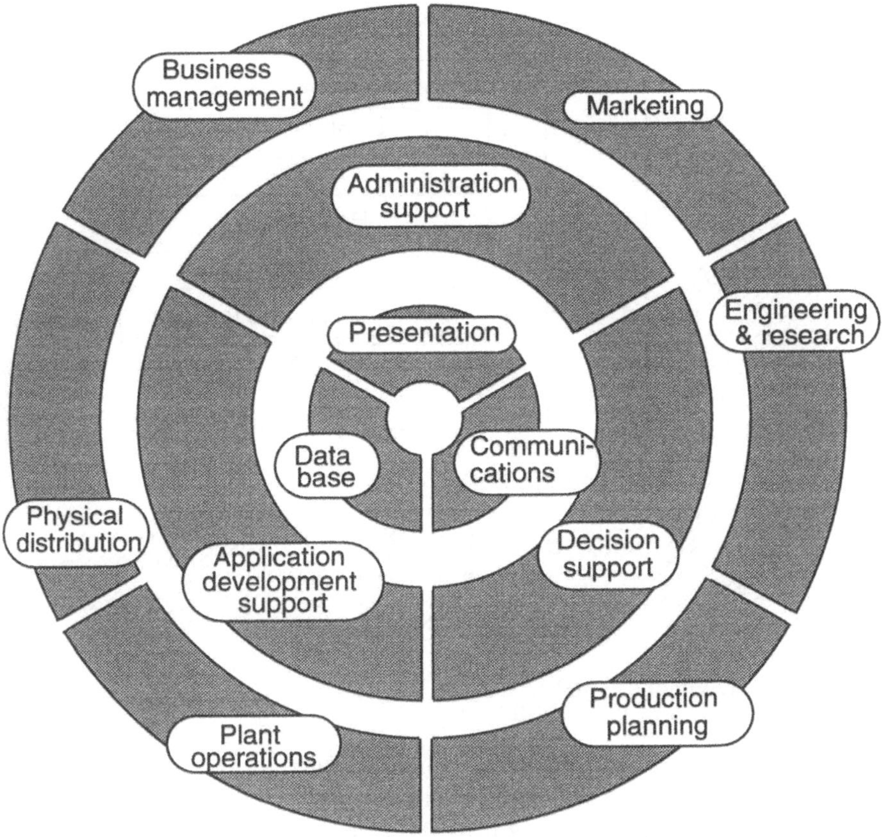

Figure 4.10 CIM architecture elements as defined by IBM.

more and more concurrent software and hardware solution for CIM coming from other IT vendors, IBM was asked by their clients to make its products compatible with the others. The approach adopted by IBM in answer to this was to offer general interfaces to integrate other products into their concepts, by using communication systems based on mainly LAN and MAP [29].

Digital has developed an Information Technology model (IT model) to support communications between two subsystems. This IT model is an extension of the OSI model, as shown in Fig. 4.11. It has adopted a layered approach.

Four layers have been defined as follows:

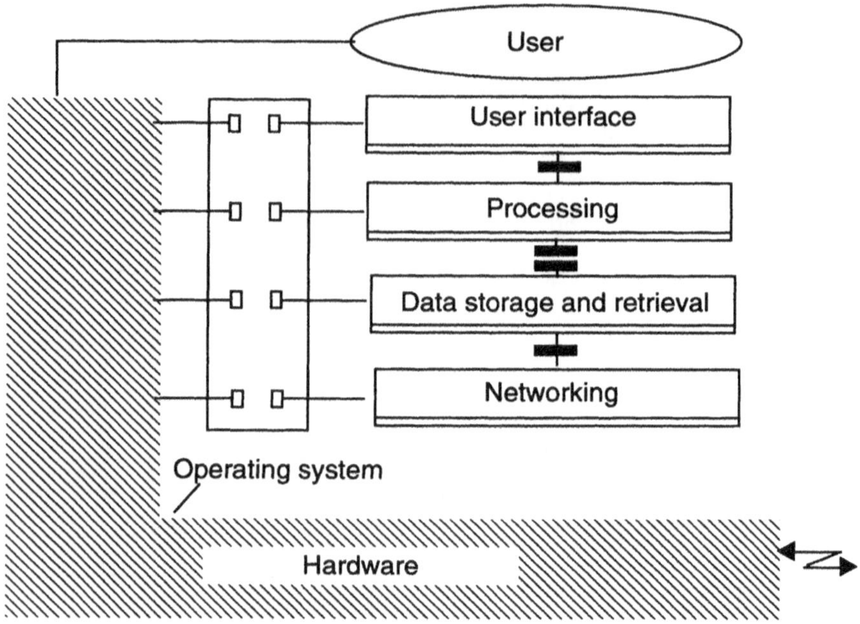

Figure 4.11 Digital IT model.

1. User (man/machine) Interface, which facilitates the transfer of data to and from the system and application;
2. Data processing, which performs the instructions requested by the user, via the user interface;
3. Data storage and retrieval, which provides services to the application to store and retrieve data;
4. Communication, which provides network links to other services and applications.

The role of the IT model is to support the interconnection of heterogeneous systems.

Another Digital CIM initiative is the NAS architecture (Network Application Support) which enables application integration and portability across a distributed, multivendor environment [29]. NAS is considered as a Digital strategy for a unified software environment allowing open and standard implementation of CIM applications.

4.7 ARCHITECTURE WORKS PERFORMED IN THE ASIA AND PACIFIC AREA

Nippon Steel Corporation reported an Industrial Automation System Model. More recently, Mitsubishi Electric Corp. has developed an object oriented approach to model factory automation systems [30]. This system architecture adopted a three level hierarchical distributed system architecture. The three levels are: factory management level or factory host-computer level; area level or FA-computer level; and Cell level or FA-controller level. As for networks, the following three kinds of networking are involved: OA-(Office Automation) networking; FA-networking; and factory backbone networking which is for interconnecting OA-networking and FA-networking. Logical modelling is based on an object-oriented approach resulting in an object oriented and class structure diagram and a message flow diagram. A case study was done on the logical modelling of the automatic assembly line control system of automobile manufacturing factories.

China launched a ten year government CIM research program in 1986 with five key reference sites for implementation. The CIM architecture and methodology projects are led respectively by the East China Institute of Technology associated with the Shenyang Automation Research Institute and the Tsinghua University. The methodology works are mainly based on an IDEF like approach. Partial implementation of IDEF0 and IDEF1x were performed at the reference sites. The architecture works are mainly based on CIMOSA. Efforts were made to develop partial models for the mechanical industry at the shop floor level. A method to create a functional model was also developed [31]. A Cooperative Program between the EEC (European Economic Community) and China in the area of CIM started in early 1992 and will run for three years. The GRAI Laboratory was chosen by the EEC as the Program Coordinator and has been involved in the methodology work.

Last but not least, the work performed at Griffith University and CSIRO Manufacturing Technology on CIM architecture and methodology should not be ignored. They consider that architecture, modelling techniques and development method are closely related. A distributed integration platform is being developed to integrate applications [32] based on an object-oriented approach. They believe that the object-oriented data models and object-oriented implementation environment can help to overcome the difficulties encountered when using a traditional top-down functional analysis and design in a heterogeneous computing environment [33].

4.8 SIMULATION MODELS

Simulation is a computer technique which consists of reproducing the dynamic behavior of a system on a computer in order to model it. This is done to get a better idea of the evolution of the system with time in a given environment and to assess its performance. Getting a good dynamic model for the simulation is difficult. Although some designers work directly with the dynamic system, one of the most often used working methods advocates a transition through a static model (see Fig. 4.12) which describes the

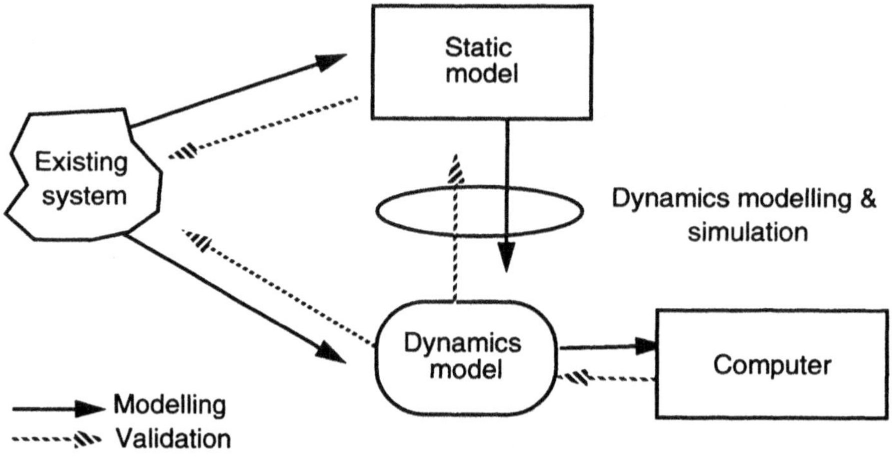

Figure 4.12 Simulation approach.

entities and their links without any notion of time (i.e. steady state). This step is essential since it allows the researcher to take into account the static aspects of the system. It also makes it possible to develop the dynamic model only when the static model has been validated. Several approaches for modelling a discrete system can be distinguished:

1. Event oriented approaches (SLAM, SIMAN etc) consist in modelling a system by describing the succession of events in the course of their evolution.
2. Activity-cycle oriented approaches (CAPS/ECSL, SIMGRAI, etc) consist in defining the activities by describing the conditions needed for their beginning and end.
3. Process-oriented approaches (SIMULA, QNAP2, etc) consist first in defining adequately the current processes, in order to cover the maxi-

mum number of current configurations. Second, it will then be enough to assemble all of the processes as needed for the elaboration of a given model.

4. Object-oriented approaches (OASYS, HOCUS, etc) consist in modelling the system by means of a set of objects which communicate between themselves by sending appropriate messages. This approach allows one to reduce the life-cycle to the steps of modelling, simulating and analysis of the results [34].

4.9 OBJECT-ORIENTED APPROACHES FOR CIM

An object-oriented approach was initially developed at IPK Berlin [35] as a programming approach. Later on it was extended to the modelling of information systems. Recently, some further work has been performed to use an object oriented approach for the modelling, the analysis and design of global CIM systems. Early work such a MOOD (Methodology for Object Oriented Design for CIM systems) was performed jointly by Digital Munich and IPK Berlin.

MOOD (see Fig. 4.13) allows the integration of two views (planner and designer). The planner and designer will be guided in the way used for mapping their objects onto a uniform object frame. The object oriented paradigm considers a system as a collection of objects and their relationships. MOOD will provide the developer with a set of generic object classes. These classes are suitable for the modelling of arbitrary CIM systems, production activity systems, etc.

As for the advantages of the object oriented approach, they consider that existing approaches (in particular, functional design and structured analysis) are partial life-cycle methods, i.e. they either focus on the analysis or on the design of the system to be developed. The requirements captured in the analysis phase cannot be mapped directly into software design. There is a 'semantic gap' between what users require and what they get. Software development requires, independent of the applied approach, that the units of the problem domain are mapped onto a software structure. In the object oriented approach, this mapping is considered to be very natural.

More recently, an IEM-GAM model (Integrated Enterprise Modelling – Generic Activity Model) was developed at IPK Berlin. The IEM approach distinguishes the objects of the manufacturing enterprise by their intended purpose into: (1) products, (2) orders and (3) resources. The kernel of the enterprise model is derived from the definition of the classes of the objects in relation to their process description with the generic activity model (IEM-

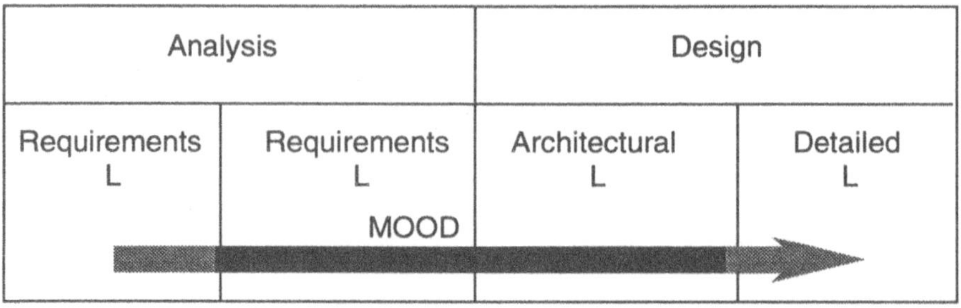

Figure 4.13 MOOD Approach.

GAM). It contains the description of two main views of the manufacturing enterprise, (i) the function model and (ii) the information model. Both views are interlinked by using the same objects and activities but looking upon and representing them in different views, ways, grades of detail, and context [35].

4.10 RECENT METHODOLOGY DEVELOPMENT AT THE GRAI LABORATORY

The ECOGRAI method developed in the GRAI Laboratory allows an investigation to design and implement a set of performance indicators for the PMS. In fact, an indicator is, first of all, a piece of information, but a special type of information, of which decision makers want to control the value. An indicator should be related to a set of decision variables which are the means for a decision maker to influence the evolution of that indicator. Implementing indicators leads to the specification of a set of potential answers for the decision makers of the production management system.

ECOGRAI uses a structured approach of four phases. Phase 1 consists of the decomposition of the project objectives. The global objectives of the production system will be decomposed into the objectives of the individual functions. Phase 2 consists of analysing the management system structure. In this phase, one identifies economic decision centers and the objectives and decision variables of each decision center. An economic decision center is a decision center which itself has a significant effect on cost. Phase 3 consists of building a set of control panels with various performance indicators displayed. The consistency between these indicators and the objectives/ decision variables must be checked at this phase. One also checks the consistency between different functions and within each function. Phase 4 is the specifications of the information system, e.g. identification of the data and data processing needed by the indicators. ECOGRAI was successfully applied in the SNECMA factory in France, and there are now other implementations.

Another work performed at the GRAI laboratory was the extension of the GRAI method to GRAI-GIM (GRAI Integrated Methodology) within the framework of the ESPRIT Project 418, Open CAM system and ESPRIT Project 2338, IMPACS [36,37]. GRAI-GIM contains a user-oriented method and a technically-oriented one (See Fig. 4.14). The user-oriented method

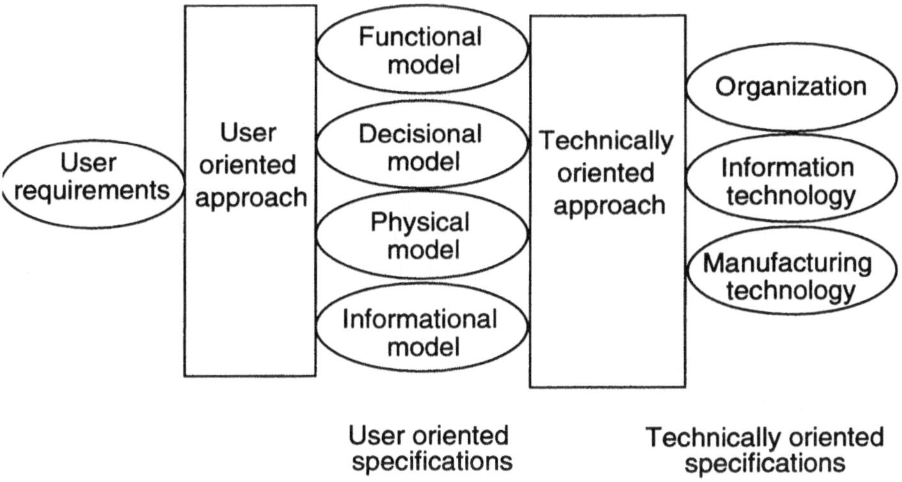

Figure 4.14 The GRAI-GIM methodology.

transforms user requirements into user specification in terms of function, information, decisions and resources. The technically-oriented method transforms the user specification into technical specifications in terms of information and manufacturing technology components and the organization. The technical specification must allow the implementor to choose (buy, commission, or develop) all the components needed to implement the system. A computerized support tool known as CAGIM (Computer Aided GIM) is being developed at the GRAI Laboratory within the framework of the IMPACS project on Unix systems with X-Windows, to support the GRAI-GIM method.

4.11 SUMMARY

Generally speaking, most of the work discussed here has concentrated on models and architectures. A few projects deal with methodologies. These were particularly IDEF and the GRAI based work. Models and methods developed in the 70s and early 80s (ICAM, CAM-I, NBS, SADT, SSAD, etc) are being challenged by new models, architectures and methods currently being developed within the ESPRIT Programme and in the USA. Particularly, CIMOSA has the most potential importance as it is striving to be the future generally-accepted CIM architecture. GRAI and ECOGRAI methods allow the user to study the decisional and economical aspects of CIM systems. GRAI-GIM (GRAI Integrated Method) provides an integrated methodology for the analysis and design of decisional, physical and information systems. Object oriented approaches currently developed at IPK Berlin have opened a new way to deal with CIM systems design. Simulation will be more and more involved in the global CIM system design. The work performed within the IFAC/IFIP Task Force has started to evaluate these major CIM models, architectures and methods, and will identify the generic requirements for a CIM architecture. All existing architectures and models are found to be heterogeneous, and the general impression is that the lack of a universally accepted CIM reference architecture can easily become the bottleneck in CIM industrial applications. System openness, portability and interoperability will progressively become a reality only through worldwide standardization.

4.12 BIBLIOGRAPHY AND REFERENCES

[1] Doumeingts, Guy, ‘Modelling Techniques for CIM’, Proceeding, CEC Workshop on Computer Aided Manufacturing, Open Systems Architecture, Brussels, Belgium (March 7–8, 1990).

[2] Chen, D, Vallespir, B., and Doumeingts, G., ‘An Integrated CIM Architecture, A Proposal’, Proceedings of the CIMCON '90, NIST Special Publication 785, National Institute of Standards and Technology, Gaithersburg, Maryland, pp 153–165 (May 1990).

[3] Hamacher, B., ‘Human Aspects of CIM’, Seminar on State of the Art in CIM, GRAI Laboratory, University of Bordeaux 1, Bordeaux, France (July 1991).

[4] Softech, Inc., Integrated Computer Aided Manufacturing (ICAM) Architecture, Part II: Composite Functional Model of 'Manufacture Product (MFGO)', Softech Inc., Waltham, Massachusetts (June 1981).

[5] CAM-I, Functional Specification for an Advanced Factory Management System, Report R-79-JSIG-01, CAM-I Incorporated, Arlington, Texas (March 1979).
[6] Jones, Albert, and Barkmeyer, Edward, 'Toward a Global Architecture for Computer Integrated Manufacturing', Proceedings of the CIMCON '90, NIST Special Publication 785, National Institute of Standards and Technology, Gaithersburg, Maryland, pp 1–20 (May 1990).
[7] Doumeingts, Guy, Methode GRAI: Methode de Conception des Systemes de Productigue, These d Etat en Automatigue, Universite de Bordeaux 1, Bordeaux, France (November 1984).
[8] Thomas, M., 'Functional Decomposition: SADT', INFOTECH State of the Art Report, Structures Analysis and Design, INFOTECH Limited, Maidenhead, Berkshire, England (1978).
[9] Gane, Chris, and Sarson, Trish, Structured Systems Analysis: Tools and Techniques, Prentice Hall, Inc., Englewood Cliffs, New Jersey (1979).
[10] De Marco, Tom, Structured Analysis and System Specification, Yourdon Press, New York, N.Y. (1978).
[11] Tardieu, Hubert; Rockfeld, Arnold; and Collette, Rene, La Methode MERISE, Principes et Outils, Les Editione d' Organization, Paris (1983).
[12] Falster, P., Rolstadas, A., and Wortmann, A., 'Factory of the Future Production Theory', FOF WP2 Report, Design of a Conceptual Model, Technical University of Norway, Trondheim, Norway (January 1991).
[13] PROCOS A/S, Functional Model for Shop and Cell Controllers, ESPRIT Project 418, Open CAM Systems, PROCOS A/S, Birkeroed Denmark (November 24, 1988).
[14] Browne, J., Harken, J., and Shivman, J., Production Management Systems, A CIM Perspective, Addison-Wesley Publishing Company, Reading, Massachusetts (1988).
[15] IMPACS Project Consortium – ESPRIT Project IMPACS, Technical Annex, ESPRIT, CEC, Brussels, Belgium (January 23, 1989).
[16] Chen, D., and Doumeingts, G., 'CIM Architecture: An Application in the ESPRIT IMPACS Project', Proceedings, China International Workshops on CIM '91, Beijing, People's Republic of China (May 7–10, 1991).
[17] AMICE Consortium, Open System Architecture for CIM, Research Reports of ESPRIT Project 688, Volume 1, Springer Verlag, Berlin (1989).
[18] Boykin, III, R.E., 'CAM-I, CIM Reference Model', Proceedings of the CIMCON '90, NIST Special Publication 785, National Institute of Standards and Technology, Gaithersburg, Maryland, pp 35–41 (May 1990).
[19] Litt, E.E., 'The Development of a CIM Architecture for the RAMP Program', Proceedings of the CIMCON '90, NIST Special Publication 785,

National Institute of Standards and Technology, Gaithersburg, Maryland, pp 251–265 (May 1990).

[20] Jung, D.W., 'Implementation of the RAMP Architecture at an Established Site', Proceedings of the CIMCON '90, NIST Special Publication 785, National Institute of Standards and Technology, Gaithersburg, Maryland, pp 266–286 (May 1990).

[21] Biemans, F.P.M. and Vissers, C.A., 'A Systems Theoretic View of Computer Integrated Manufacturing', Proceedings of the CIMCON '90 NIST Special Publication 785, National Institute of Standards and Technology, Gaithersburg, Maryland, pp 390–410 (May 1990).

[22] Yoder, J.R., 'Toward a New CIM Architecture for SANDIA Laboratories', Proceedings of the CIMCON '90, NIST Special Publication 785, National Institute of Standards and Technology, Gaithersburg, Maryland (May 1990).

[23] Anderson, Allan; Jenne, Theresa and Mikkilineni, Krishna, 'CIM Architecture, One Perspective'. Proceedings of the CIMCON '90, NIST Special Publication 785, National Institute of Standards and Technology, Gaithersburg, Maryland, pp 506–524 (May 1990).

[24] Williams, T.J., Editor, A Reference Model for Computer Integrated Manufacturing (CIM). A Description From the Viewpoint of Industrial Automation, CIM Reference Model Committee, International Purdue Workshop on Industrial Computer Systems, Instrument Society of America, Research Triangle Park, North Carolina (1989).

[25] Williams, T.J., The Purdue Enterprise Reference Architecture, Instrument Society of America, Research Triangle Park, North Carolina (1992).

[26] Roboam, M. and Fox, M.S., 'Intelligent Networking: How A.I. Can Help Integrating the Manufacturing Enterprise,' Proceedings CIM '90 Conference, Bordeaux, France (June 1990).

[27] Roboam, M., Sykara, K., and Fox, M.S., 'Organization Modelling as a Platform for Multiagent Systems', Proceedings, Computer Applications in Production and Engineering, CAPE '91, Doumeingts, G., Browne, J. and Tomlyanovich, M., Editors, North Holland, Amsterdam, pp 581–591 (1991).

[28] Rembold, Ulrich, and Nannaji, B., 'The role of Manufacturing Models for the Information Technology of the Factory of the 1990's', Journal of Design and Manufacturing, Vol. 1, pp 67–87 (1991).

[29] Anonymous, NAS (Network Application Support) Handbook: Developing Applications in a Multivendor Environment, Digital Equipment Corporation, Marlboro, Massachusetts (1990).

[30] Inamoto, Atsushi, and Shirakura, Tadaaki, 'Modelling of a Factory Automation System by an Object Oriented Approach', Minutes, Workshop Meeting, IFAC/IFIP Task Force on Architectures for Integrating Manufacturing Activities and Enterprises, Bordeaux, France (September 5–7, 1991).
[31] Wang, K.H., and Deng, Z.Q., 'Models and Methodology Study for CIMS Function Structure Design', Proceedings, Computer Applications in Production and Engineering, CAPE '91, Doumeingts, G., Browne, J., and Tomlyanovich, M., Editors, North Holland Amsterdam, pp 325–332 (1991).
[32] Bernus, Peter, 'Comparative Analysis of CIM Models and Architecture', Minutes Workshop Meeting, IFAC/IFIP Task Force on Architecture for Integrating Manufacturing Activities and Enterprises, Bordeaux, France (September 5–7, 1991).
[33] Bernus, Peter, 'Object-Oriented Glue to Stick Pieces of CIM Together', Proceedings CIM90 Conference, Bordeaux, France (June 12–14, 1990).
[34] VALORIS, OASYS: Outil Pour la Mise en Oeuvre de Simulation d'Un Systime de Production Raport AD1 VALORIS, SGN, Paris (June 1986).
[35] Martins, K. Sussenguth, W., and Jochem, R., 'Integrated Information Modelling for CIM: An Object Oriented Method for Integrated Enterprise Modelling', Proceedings Computer Applications in Production and Engineering, CAPE '91, Doumeingts, G., Browne, J., and Tomlyanovich, M., Editors, North Holland, Amsterdam, The Netherlands, pp 315–323 (1991).
[36] Vellespir, B., Doumeingts, G., and Zanettin, M., 'Proposal for an Integrated Approach to Model and Design Manufacturing Systems', Proceedings, Third International Conference on Computer Application in Production and Engineering, CAPE '89, Tokyo, Japan (October 2–5, 1989). Published by North Holland.
[37] Zanettin, M., and Doumeingts, G., 'The GIM Method for CIM System Analysis', for Advanced Studies in Systems Research and Cybernetics, International Institute for Advanced Studies in Systems Research and Cybernetics, Baden Baden, Germany (August 1992).

4.13 FURTHER READING

Doumeingts, Guy, 'How to Decentralize Decisions Through GRAI Model in Production Management', *Computers in Industry*, **6** (6) (1985).

Doumeingts, Guy, Poumeyrol, E., and Roboam, M., GRAI Analysis Report, ESPRIT Project 418, Open CAM System, Technical Report, GRAI Laboratory, University of Bordeaux 1, Bordeaux, France (May–November 1986).

5

Some important articles about enterprise integration

T.J. Williams

The first ten models or architectures (the labels are synonymous in this context) are grouped in approximately their order of importance as judged by this author. The remainder are listed simply in alphabetical order.

The AMRF model of the U. S. National Institute of Standards and Technology (NIST) (formerly National Bureau of Standards (NBS) is listed first because it was the first model published as a CIM architecture and has been widely accepted.

CIMOSA is listed second because of its major political importance due to its backing by the European Communities and their push for its acceptance as a standard. The ISO TC184/SC5/WG1 group has the mandate from the International Standards Organization to recommend standards in this area and – as of 1992-93 – has its own model to promote in competition with CIMOSA. GRAI-GIM is listed next in its order in the evaluation by the Task Force for this report.

The Purdue University model is the oldest of the reference models but was labelled as Hierarchical Control or Millwide Control during the 1970s and was only recently identified with the title Computer Integrated Manufacturing. This has been due to its process industry application context.

The Case–Western Reserve model is the oldest of all but was confined mainly to its use as a demonstration vehicle for advanced control theory topics. The Computer Integrated Manufacturing International (CAM-I) Model was a precursor to the ISO offering. The Nippon Steel model was the first major attempt to incorporate enterprise requirements in such a model. The Society of Manufacturing Engineers and IBM have similar models to each other.

The book by Thacker ([1], Part II of Section 5.9), published by the Society of Manufacturing Engineers, presents a very interesting and descriptive model of CIM from the viewpoint of the CEO of a company interested in the technology. It also presents a quite thorough discussion (description only) of the 'life-cycle' process of CIM system development. However it goes into none of the detail and technology involved in the three references architectures evaluated in this report. It therefore was not included among them, although it is an excellent semitechnical reference to supplement any of them.

5.1 NIST

United States National Institute of Standards and Technology (NIST), formerly National Bureau of Standards (NBS).

[1] Albus, J., Barbera, A., and Nagel, N., 'Theory and Practice of Hierarchical Control,' Proceedings of the 23rd IEEE Computer Society International Conference, pp. 18–39 (September 1981).
[2] Simpson, J. A., Hocken, R. J., and Albus, J. S., 'The Automated Manufacturing Research Facility of the National Bureau of Standards,' Journal of Manufacturing Systems, Vol. 1, No. 1, pp. 17–32 (1982).
[3] Jones, A and Whitt, N., Eds., Proceedings on Factory Standards Model Conference, National Bureau of Standards, Gaithersburg, Maryland (1985).
[4] Jones, A. and McLean, C., 'A Proposed Hierarchical Control Model for Automated Manufacturing Systems,' Journal of Manufacturing Systems, Vol. 5, pp. 15–25 (1986).
[5] Jones, A. Barkmeyer, E. and Davis, W., 'Issues in the Design and Implementation of a System Architecture for Computer Integrated Manufacturing,' Int. J. Computer Integrated Manufacturing, Vol. 2, No. 2, pp. 65–76 (March–April 1989).

5.2 CIMOSA

[1] ESPRIT, Project 688, AMICE, OPen Systems Architecture for CIM, Springer Verlag, Belin (1988)
[2] ESPRIT, Project 688, CIMOSA Reference Architecture Specification, ESPRIT Consortium AMICE, Brussels, Belgium (May 1989).
[3] ESPRIT Consortium AMICE, Open System Architecture, CIMOSA, AD 1.0, Architecture Description, ESPRIT Consortium AMICE, Brussels, Belgium (1991).
[4] Anonymous, AMICE: CIMOSA, ESPRIT Project 5288, Milestone M-2, AD 2.0, Volume 2, Architecture Description, Document R0443/1, Consortium AMICE, Brussels, Belgium, (August 24, 1992).

[5] Beeckman, Dirk, 'CIMOSA: Computer Integrated Manufacturing Open Systems Architecture,' Int. J. Computer Integrated Manufacturing, Vol.. 2, No. 2, pp. 94–105 (March–April 1989).
[6] Jorysz, H. R. and Vernadat, F. B, 'CIMOSA Part 1: Total Enterprise Modelling and Function View,' Int. J. Computer Integrated Manufacturing, Vol. 3, Nos. 3 and 4, pp. 144–156 (1990).
[7] Jorysz, H. R. and Vernadat, F. B., 'CIMOSA Part 2: Information View,' Int. J. ComputerIntegrated Manufacturing, Vol. 3, Nos 3 and 4, pp. 157–167 (1990).
[8] Klittich, M., 'CIMOSA Part 3: CIMOSA Integrating Infrastructure – The Operational Basis for Integrated Manufacturing Systems,' Int. J. Computer Integrated Manufacturing, Vol. 3, Nos. 3 and 4, pp. 168–180 (1990).

5.3 ISO/TC–184/SC5/WG1 REFERENCE MODEL FOR COMPUTER INTEGRATED MANUFACTURING

[1] ISO (International Organization for Standardization), ISO TC184/SC5/WG1, 'The Ottawa Report on Reference Models for Manufacturing Standards,' Report N51 (September 14, 1986).
[2] 2.Thomson, V. and Graefe, U., '1CIM – A Manufacturing Paradigm, SME MM86-722,' 5th Canadian CAD/CAM and Robotics Conference, Toronto (1986).
[3] .CIM Reference Model Working Group, ISO/TC184/SC5/WG1, Technical Report on CIM Reference Model (July 1987).
[4] ISO (International Organization for Standardization), 'Reference Model for Shop Floor Standards, Part 1,' ISO TC184/SC5 N131 (July 15, 1988).
[5] Graefe, U. and Thomson, V. 'A Reference Model for Production Control,' Int. J. Computer Integrated Manufacturing , Vol. 2, No. 2, pp. 86–93 (March–April 1989).
[6] ISO/TC184/SC5/WG1, 'Reference Model for Shop Floor Production Standards, Part 1: A Reference Model for Standardization and a Methodology for Identification of Standards Requirements,' Technical Report (1989).

5.4 GRAI LABORATORY, UNIVERSITY OF BORDEAUX

[1] Doumeingts, G., Methode de Conception des Systemes en Productique, Thesis d'etat en Automatique, Universite de Bordeaux I, Bordeaux, France (November 13, 1984).

[2] Doumeingts, G., Vallespir, B. Darracar, d. and Roboam, M., 'Design Methodology for Advanced Manufacturing Systems,' Computers in Industry, Vol. 9, No. 4, pp. 271–296 (December 1987).
[3] Doumeingts, G., Vallespir, B., Zanettin, M., and Chen, D., GIM, GRAI INTEGRATED METHODOLOGY, A Methodology for Designing CIM Systems, Version 1.0, Unnumbered Report, LAP/GRAI, University Bordeaux 1, Bordeaux, France (May 1992).

5.5 PURDUE LABORATORY FOR APPLIED INDUSTRIAL CONTROL, PURDUE UNIVERSITY

[1] Williams, T. J., 'Computer Control in the Steel Industry,' Computers in Mechanical Engineering, Vol. 2, No. 4, pp. 14–26 (January 1984).
[2] Project Staff, Tasks and functional Specifications of the Steel Plant Hierarchical Control System, Report Number 98, Purdue Laboratory for Applied Industrial Control, Purdue University, West Lafayette, IN (September 1977; Revised June 198). Published in Book Form as: Williams, T. J., Editor, Analysis and Design of Hierarchical Control Systems, Elsevier Science Publishers, Amsterdam, The Netherlands (1985).
[3] Brewster, D. B., Uronen, P. and Williams, T. J., Hierarchical Computer Control in the Paper Industry, Report Number 111, Purdue Laboratory for Applied Industrial Control, Purdue University, West Lafayette, IN (March 1985).
[4] Systems Engineering Group, INCOS Project, Tasks and Functional Specifications of the Bhilai Steel Plant Integrated Control System (INCOS), Steel Authority of India, Ltd., Delhi, India (April 1986).
[5] McCarthy, J. J and Ruckman, R. P., 'The Application of the CIM Reference Model to a Continuous Process Plant,' Advanced Control in Computer Integrated Manufacturing, Morris, H. M., Kompass, E. J. and Williams, T. J., Editors, Purdue University, West Lafayette, IN, pp. 87–96 (September 28–30, 1987).
[6] Williams, T. J., Editor, A Reference Model for Computer Integrated Manufacturing (CIM), A Description From the Viewpoint of Industrial Automation, Minutes, CIM Reference Model Committee, International Purdue Workshop on Industrial Computer Systems, Purdue University, West Lafayette, IN (1988), Instrument Society of America, Research Triangle Park, NC (1989).
[7] Williams, T. J., 'A Reference Model for Computer Integrated Manufacturing From the Viewpoint of Industrial Automation,' International Journal

of Computer Integrated Manufacturing, Vol. 2, No. 2, pp. 114–127 (March–April 1989).
[8] William, T. J., The Purdue Enterprise Reference Architecture, Report Number 154, Purdue Laboratory for Applied Industrial Control, Purdue University, West Lafayette, IN (1991). Also published Instrument Society of America, Research Triangle Park, North Carolina (1992).

5.6 SYSTEMS RESEARCH CENTER, CASE–WESTERN RESERVE UNIVERSITY

[1] Findeisen, W. and Lefkowitz, I., 'Design and Applications of Multilayer Control,' Proc. Fourth IFAC Congress, Warsaw, Poland (1969).
[2] Mesarovic, M. D., 'Multilevel Systems and Concepts in Process Control,' Proc. of IEEE, 58, 111–125 (1970).
[3] Mesarovic, M. D., Macko, D. and Takahara, Y., Theory of Hierarchical Multilevel Systems, Academic Press, New York (1970).
[4] Schoeffler, J. D., 'Multilevel Systems and Decomposition for the Solution of Static Optimization Problems; Decomposition and Multilevel Methods of On Line Computer Control of Industrial Processes,' In Wismer, D. A., Ed., Optimization Methods for Large Scale Systems, McGraw Hill, New York (1970).

5.7 COMPUTER-AIDED MANUFACTURING INTERNATIONAL

[1] Anonymous, Functional Specification for An Advanced Factory Management System, Report R-79-JSIG-01, CAM-I Incorporated, Arlington, Texas (March 1979).
[2] Boykin, III, R.E., 'CAM-I, CIM Reference Model', Proceedings of the CIMCON '90, NIST Special Publication 785, National Institute of Standards and Technolgy, Gaithersburg, Maryland, pp 35–41 (May 1990).
[3] Throop, J. W. and Read, R. R., A Discrete Parts Manufacturing Model, Report Number R-83-SC-01, Computer-aided Manufacturing International, Inc., Arlington, Texas (August 1983).
[4] Lui, J., 'The CAM-I Advanced Factory Automation System,' Proceedings on Factory Standards Model Conference, National Bureau of Standards, Gaithersburg, Maryland (1985).

5.8 NIPPON STEEL CORPORATION

[1] Anonymous, Industrial Automation System Model, Nagoya Works, Nippon Steel Corporation, Nagoya, Japan (June 11, 1987).
[2] See Appendix V in [6], Section 5.5, of this bibliography for the discussion and presentation of this model.

5.9 SOCIETY OF MANUFACTURING ENGINEERS (SME)

Part I

[1] Bertain, L. and Hales, L., Eds., A Program Guide for CIM Implementation, Second Edition, Society of Manufacturing Engineers, Dearborn, MI (1987).
[2] Bertai, L., CIM Implementation Guide, Third Edition, Computer and Automation Systems Association, Society of Manufacturing Engineers, Dearborn, MI (1991).

Part II

[1] Thacker, R. M., A Blueprint for the Computer Integrated Manufacturing Enterprise, Society of Manufacturing Engineers, Dearborn, MI (1989).

5.10 INTERNATIONAL BUSINESS MACHINES CORPORATION (IBM)

Part I

[1] Anonymous, Computer Integrated Manufacturing, The CIM Enterprise, Document G320-9802-00, International Business Machines Corporation, White Plains, NY (April 1989).
[2] Anonymous, IBM CIM Architecture, Document G520-6686-00, International Business Machines Corporation, White Plains, NY (October 1989).

Part II

[1] Bachman, J. A., 'A Framework for Information Systems Architecture,' IBM Systems Journal, Vol 26, No. 3, pp. 276–292 (1987).
[2] Zachman, J. A. Application and Data: Factory of the Future, Applications Technology Group, Information Systems Department, International Business Machines Corporation, Los Angeles, CA (May 23, 1989).

5.11 AMERICAN SOCIETY FOR TESTING AND MATERIALS (ASTM)

[1] Subcommittee E31.20, Computerized Manufacturing, of committee E-31, Computerized Systems, American Society for Testing and Materials, 'Standard Guide for Procurement of Computer-Integrated Manufacturing Systems,' Standard E 1383–89, Annual Book of ATM Standards, Vol. 14.01(Approved January 27, 1989, Published March 1989).

5.12 CARNEGIE–MELLON UNIVERSITY

[1] Beck, C. L., Modelling and Simulation of Flexible Control Structures for Automated Manufacturing Systems, Technical Report, Robotics Institute, Carnegie–Mellon University (1985).

5.13 DIGITAL EQUIPMENT CORPORATION (DEC)

[1] Flateau, U., Digital's CIM Architecture, Revision 1.1, Digital Equipment Corporation, Marlboro, MA (April 1986).
[2] Digital Equipment Corporation, Proceedings, CIM Management Forum for the Process Manufacturing Industries, Orlando, Florida (January 25–27, 1988).

5.14 HONEYWELL, INC.

[1] McCarthy, J. J. and Mikkilineni, K., 'Considerations for Distributed Data Bases in the Computer Integrated Business,' Paper presented at the 43rd Annual Conference and Exhibition, Instrument Society of America, Houston, TX (October 16–21, 1988).
[2] Mikkilineni, K., Clark, B., Jenne, T., and Krueger, K., HCAM Architecture 1988 Final Report, Vol. 1–4, Honeywell (1988).
[3] Anderson, A., Jenne, T. Mikkilineni, K., HCAM Architecture 1989 Final Report, Vol. 1 and 2, Honeywell, (1989).
[4] Anderson, A., Jenne, T. and Mikkilineni, K., 'CIM Architecture: One Perspective,' Honeywell, Inc., Minneapolis, MN (1990). Also Proceedings of CIMCON '90, NIST Special Publication 785, National Institute of Standards and Technology, Gaithersburg, Maryland, pp 506–524 (May 1990).

5.15 INDUSTRIAL TECHNOLOGY INSTITUTE (ITI)

[1] Parunak, H.V.D., White, J. F., Judd, R., Irish, B.W. and Kendrick, J., An Architecture for Heuristic Factory Control, Industrial Technology Institute, Ann Arbor, MI (March 26, 1986).
[2] Parunak, H.V., and White, J.F., 'A Framework for Comparing Factory Reference Models', 1987 IEEE Workshop on Languages for Automation, Vienna, Austria, (August 24–27, 1987).
[3] Judd, R.P., Vanderbik, R.S., Brown, M.S., and Sauter, J.A., 'Manufacturing System Design Methodology: Execute the Specifications (X Spec)', Proceedings of CIMCON '90 NIST Special Publication 785, National Institute of Standards and Technology, Gaithersburg, Maryland, pp 133–152, (May 1990).

5.16 THE PROCOS–A/S SYSTEM

[1] Moss, S. P., 'A Management and Control Architecture for Factory-Floor Systems: From Concept to Realty,' Int. J. Computer Integrated Manufacturing, Vol. 2, No. 2, pp. 106–113 (March–April 1989).

5.17 SI GROUPS – LOUGHBOROUGH UNIVERSITY, UK

[1] Weston, R. H., Hodgson, A., Gascoigne, J. D., Sumpter, C. M., Rui, A. and Coutls, I., 'Configuration Methods and Tools for Manufacturing Systems Integration,' Int. J. Computer Integrated Manufacturing, Vol. 2, No. 2, pp. 77–85 (March–April 1989).

5.18 SOFTECH, INC. (ICAM)

[1] SOFTECH, Inc., Integrated Computer-Aided Manufacturing (ICAM), Task I – Final Report, Manufacturing Architecture, Air Force Materials Laboratory, Wright Patterson Air Force Base, OH (November 1978).

5.19 UNIVERSITY OF TWENTE, THE NETHERLANDS

[1] Biemans, F. P. M., A Reference Model for Manufacturing Planning and Control, Doctoral Dissertation, University of Twente, Enschede, The Neth-

erlands (October 27, 1989). Published by: CID–Gigevens Koninklijke Bibliotheek, Den Haag, The Netherlands (1989).
[2] Biemans, F.P.M. and Vissers, C.A., 'A Systems Theoretic View of Computer Integrated Manufacturing', Proceedings of the CIMCON '90 NIST Special Publication 785, National Institute, Standards and Technology Gaithersburg, Maryland, pp 390–410 (May 1990).

5.20 SOME U.S. DEPARTMENT OF DEFENSE MODELS AND ARCHITECTURES FOR WHICH THIS WRITER HAS NO DETAILS AT PRESENT

1. Honeywell's EIS (Engineering Information System)
2. Product Data Exchange Specification (PDES) by PDES, Inc.
3. SEMATECH CIM Architecture
4. DARPA Initiative Into Concurrent Engineering (DICE) being developed by W Va, CMU and RPI Universities
5. The RPI CIM Model (Rensselaer Polytechnic Institute)
6. Advanced Cost Management System (DOD)
7. Microelectronics Manufacturing Science and Technology (MMST) DOD
8. CFI (DOD) (CAD Framework Initiative)
9. Lehigh University, AMEF, Agile Manufacturing Enterprise Forum. (No Specific Reference Avaliable).

The above list developed by:
Enterprise Integration Framework Project,
Federal Systems Division,
IBM Corporation, Oswego, NY 13827, USA

5.21 SOME ADDITIONAL MODELS RECOMMENDED FOR STUDY BY THE ISO WORKING GROUP, ISO/TC-184/SC5/WG1

1. Additional Models Recommended for Study by ISO/TC-184/SC5/WG1
2. EISWG (MCC) (Enterprise Integration Suppliers Working Group)
3. EIP (Enterprise Integration Program)
4. NCMS/CIO (National Center for Manufacturing Studies) (DFI)
5. CASEE (Hewlett–Packard)
6. OMG (Object Management Group)
7. C4 (GM)
8. PWAF (Plant With a Future – Caterpillar)

5.22 SOME PUBLICATIONS OF GENERAL IMPORTANCE IN THE STUDY OF CIM REFERENCE MODELS AND ARCHITECTURES

[1] Jones, A., and Davis W., Eds., 'Special Issue on Architectures and Implementation,' Int. J. Computer Integrated Manufacturing, Vol. 2, No. 2, London (March–April 1989). This issue contains six articles and is an excellent review of the field as of its date of publication.

[2] Merchant, M. E., Nemes, L. and Bernus, P., Special Issue, Joseph Hatvany Memorial, Computers in Industry, Vol. 14, No. 1–3 (May 1990).

[3] Milacic, V. R. and Majstorovic, V. D., Eds., Special Issue, Intelligent Manufacturing Systems – IMS'89, Computers in Industry, Vol. 15, No. 1–2, pp. 1–162 (October 1990).

[4] Panel on Technical Barriers to Computer Integration of Manufacturing, A Research Agenda for CIM, Information Technology, National Research Council, National Academy Press, Washington, D.C. (1988).

[5] Sobczak, T. V., Ed., A Glossary of Terms for Computer Integrated Manufacturing, Computer and Automated Systems Association of SME, Dearborn, MI (1984).

[6] U.S. Congress, Office of Technology Assessment, Making Things Better: Competing in Manufacturing, OTA-ITE-443, U.S. Government Printing Office, Washington, D.C. (February 1990).

6

A summary of the CIMOSA reference architecture

J. Vlietstra

6.1 CIMOSA AND ITS ASSOCIATED CONSTRUCTS [1]

The ESPRIT project AMICE[1] has developed an Open System Architecture (CIMOSA) for the manufacturing industry. The main goal of CIMOSA is to support process oriented modelling of manufacturing enterprises and to provide execution support for operation of enterprise systems based on those models. Emphasis in CIMOSA is on modelling support for business users[2] rather than for modelling specialists.

6.2 ENTERPRISE INTEGRATION – THE SOLUTION

Enterprise integration has to provide for flexibility of enterprise operations and efficient use of enterprise assets. Therefore, a modular approach to integration is the most promising one. This is an approach which models the operation as a set of cooperating processes exchanging results and requests (events) between themselves. Only the exchanged objects need to have a representation which is common between partners. This minimal unification of objects, together with an integrating infrastructure for hiding heterogeneity, can readily cope with enterprise integration. The CIMOSA approach enables integration of both reengineered and existing systems.

1. AMICE: European Computer Integrated Manufacturing Architecture (reverse acronym)
2. Business users (or business professionals) are all those people with responsibilities for particular business processes. Business users are: technical staff, administrative staff and all levels of management.

6.3 MANAGEMENT OF CHANGE

Companies are not always organized for the use of fast decision making processes. Departments are still managed according to their own subgoals rather than working toward the overall enterprise objectives. Responsibilities are still structured in one dimensional hierarchies which mix responsibilities for enterprise assets with those for enterprise operations. Matrix organizations are still a rather theoretical concept. In addition, the decision making process in many companies is still based on traditional information processing, that is, information gathering with 'paper and pencil,' on request and from inconsistent sources. This process is defined from the view of a rigid management hierarchy and is very prone to severe information losses. The process is very time consuming and often yields only insufficient or even wrong information and decisions. To achieve 'real time' operational flexibility, delegation of responsibility and authority are needed to allow people to make changes as they are required rather than going through a hierarchy of many levels for decision sign-offs. Ease of access and more efficient use of information will enable the delegation of more responsibility and authority.

Therefore, active management of change is the most significant future requirement for successful enterprise operation. This means recognizing and reacting to external changes as early as possible and defining and implementing the internal modifications to their response. Business process reengineering and simplification as tasks of enterprise engineering are prerequisites for enterprise integration.

Decision support is an essential part of the management of change. This implies a timely access to the right information and the distribution of the decision making process results to the right places. Decision making requires clear and explicit knowledge about the relevant business processes and their contents (information, resources, responsibilities and authorities), a definition of alternative solutions, and an analysis of their impact on the total enterprise operation. Therefore computer simulation of potential solutions is a requirement for future decision support systems.

CIMOSA provides decision support through its up-to-date models (identification of relevant information, analysis of alternative solutions and propagation of verified decisions) and ease of business process reengineering.

6.4 SCOPE OF ENTERPRISE INTEGRATION

Industry needs an implementation strategy in which an existing, semiautomated enterprise can steadily progress in discrete steps towards CIM. The manufacturing plant must continue to operate day by day, but, at the same time, incorporate changes in the direction of both automation and integration. Since it is impossible to purchase all of the components required for realizing an enterprise environment from a single vendor, standards are required to insure the correct interworking of components from multiple sources.

In addition to its capital investments, the enterprise's knowledge about its business and technical processes is one of its most valuable assets, often differentiating it from its competitors. The integration of this internal knowledge with the external one arising from other sources (e.g. trade associations, de facto standards, formal standards) is vital for an enterprise trying to remain competitive and cost effective.

Enterprise operation also consists of internal and external parts. Internal operation of an enterprise is that set of business processes needed to market, develop, manufacture, distribute and sell products and to administer and manage the operation itself. External operations which are coupled to the internal ones cover relationships with suppliers, customers, financial institutions, government agencies, etc. The main focus of enterprise integration is on internal environments. However, relations to the external environments and their impact on the internal operation have to be made visible as well. Only if these dependencies are known and have become part of the business model can the impact of changes be fully evaluated. However, more detailed modelling is required for internal operations than for external ones. Whereas the control flow of internal business processes is needed even for simulation of alternatives, identification of shared information and dependencies is mostly sufficient for modelling the external relations.

Information technology hardware makes it possible even today to install very large networks of computer systems with an almost unrestricted performance and processing capability. However, the ability to handle the vast amount of information needed and existing in the manufacturing enterprise and to process the right information, for the right purpose, at the right time and in the right place is still a major problem.

CIMOSA provides means to describe in a consistent way both internal and external enterprise operations and their information needs. In addition, it clearly distinguishes between model engineering and model use and defines in its system life-cycle a formal engineering release for operation.

6.5 ENTERPRISE ENGINEERING AND OPERATION

Enterprise evolution leads to a continuous need for enterprise integration. This is a need to be fulfilled by new engineering tasks concerned with maintaining and extending enterprise operation efficiency and flexibility. Therefore, enterprise integration has to be conceived as a modular approach structuring and modelling the business into manageable units (processes) which cooperate with each other according to identified needs and sharing information on request. Only with the enterprise modelled as a set of cooperating processes rather than as a large monolithic entity can changes be implemented and operational modifications and extensions be made in 'real time.' Only then can sufficient operational and organizational flexibility be achieved.

Information Technology (IT) based integration support for business model engineering provides:

1. Definition of business process flow of control and information needs.
2. Definition and organization of enterprise assets (resources and information).
3. Definition and organization of enterprise responsibilities and authorities.
4. Maintenance of business models (modifications and extensions) by the business user.

The set of business processes which make up the enterprise operation have to operate and interoperate in a highly efficient and flexible manner. To assure operational flexibility and efficiency an easily accessible knowledge of the operation has to be provided through up-to-date business models. Better structuring and modelling of the enterprise operation will hide process complexity and thereby improve decision making and business management.

Operational flexibility will be greatly enhanced by the direct use of models for operational control and monitoring. Business changes will be implemented and validated in the model and directly released for operation. Improved enterprise logistical control and superior asset management will also contribute to operational flexibility and efficiency.

IT based integration support for business model execution provides:

1. Evaluation of impact of change and alternative solutions (model simulation).
2. Model based operation control and monitoring.

6.6 COEXISTENCE WITH HERITAGE/LEGACY SYSTEMS

New system components, reengineered according to new paradigms in modelling still have to interoperate with the existing parts of the enterprise operation. With an average life time of about 12 years for Mainframe applications [2] there will be many years of such coexistence.

Therefore, any new methodology and technology used in enterprise operation has to provide means for interoperating with the existing world. The most promising approach to achieve interoperability between incompatible systems is through the use of information objects shared between different parts of the system. With this approach agreement on common representation is only necessary between partners sharing the same objects. This enables the identification of classes of information objects according to their degree of sharing in terms of private, common or public use. This is the approach taken also by people working in artificial intelligence (AI) and on the development of ontologies (generic theories of common aspects of reality).

CIMOSA provides solutions for the enterprise integration of both reengineered and heritage systems.

6.7 BUSINESS BENEFITS

Enterprise reengineering and process simplification are tasks which will coordinate enterprise evolution as well as improving enterprise performance in general. In the IT arena the availability and sharing of information between different business areas as well as the common use of computer services will reduce operations costs, but more importantly it will improve operational quality and flexibility. Providing relevant information in real time to the decision makers will enable a faster reaction to different market changes (markets on goods, services, knowledge, technology and money). It will also help to base enterprise strategies on real life facts rather than fictions.

CIMOSA will provide benefits for enterprises by:

1. Improving enterprise operational flexibility and efficiency by the reengineering and simplification of business processes.
2. Supporting the management of change by evaluation of alternatives through the simulation of operations.
3. Improving operational flexibility and efficiency and reducing operational costs through better business management (people, process, resources, information).

4. Reducing lead times through the sharing and reuse of relevant information, modelling building blocks and system components.

6.8 ENTERPRISE MODELLING AND ITS REQUIREMENTS

Enterprise business modelling is a prerequisite for successful enterprise integration [3]. However, enterprise operations have to be well understood and explicitly described and presented in order to identify inconsistencies and incompatibilities and to analyze their consequences. Alternative solutions can then be modelled and evaluated through simulation. Also enterprise modeling has to meet the requirements of a number of different users in their day-to-day operations. Therefore modelling has to be based on business objectives and must describe the operation in terms of its related functionality and dynamic behavior (control flow).

Enterprise modelling should not be done as a one time all encompassing venture. A modular structure will allow for evolutionary model building and model maintenance. However, to assure consistency, all modules have to be parts of, derived from and linked to a common model. These submodels must meet specific user needs for optimising and structuring certain aspects of the operation without being constrained by a huge model. Levels of abstraction are needed to support strategic, tactical and operational planning and decision making. Again, all levels have to be abstractions from the same underlying model.

Model engineering needs heavy IT support to enable the creation of consistent and easily maintainable and extendible models. The users have to be guided through the model engineering process by providing them with an identification of those reusable building blocks already known to the system. Ease of maintenance is needed to adapt models in real time to the changing internal and external environment as well as to enable the evaluation of alternatives to the existing situation ('as-is' versus 'to-be' models).

Unlike other approaches CIMOSA allows the users to model different parts of the enterprise separately and to integrate them later. CIMOSA provides a modular approach (domain process) for business modelling by identifying three modelling levels (requirements definition, design specification and implementation description) and four views (function, information, resource, organization) as part of an open set.

6.9 AMICE PROJECT

The goal of the AMICE project is to develop an Open System Architecture for CIM (CIMOSA) which will facilitate continuous enterprise evolution and make needed operational improvements manageable while at the same time providing a strategy to protect current and planned investment in an enterprise's asset base. Such an architecture will become the base for consistent and maintainable enterprise engineering and operation support and thereby be the cornerstone for an industrial infrastructure for industry wide use of information and supporting technology.

CIMOSA provides a consistent enterprise model building and maintenance methodology and an integrating infrastructure for model engineering and execution support. CIMOSA is based on a process oriented modelling approach describing all enterprise activities in a common way. Such activities include manufacturing processes on the shop floor as well as management and administrative processes.

Validation of the concepts has been carried out in various case studies in consortium member companies. The work has led to inputs into the standardization of the modelling and integrating infrastructure. The latter takes into account current work on standards like OSI (Open System Interconnection), ODP (Open Distributed Processing), etc. Standardization has been achieved for the Modelling Framework by a European PreNorm [4] which has become the base for the work on international standardization in ISO TC 184/SC5/WG1. Project results are summarized in 'Open System Architecture for CIM' [5], several publications by project members [6] and a recent private publication by the ESPRIT Consortium AMICE [7].

CIMOSA is a prenormative development effort aimed at guiding and supporting users and vendors in enterprise integration as well as in the development of compliant products which will ease enterprise integration. It will be usable in an evolutionary mode starting with enterprise model engineering and decision support and moving gradually into new modes of enterprise operation and management. To further verify CIMOSA goals and objectives the project concentrates on feasibility studies of the implementations of CIMOSA models and the Integrating Infrastructure which prepare the project for a future implementation of an industrial pilot.

6.10 CIMOSA APPROACH

CIMOSA provides a framework for guiding CIM users in enterprise system design and implementation, and CIM vendors in system component development. It provides a descriptive, rather than a prescriptive methodology supporting the system life-cycle. CIMOSA does not attempt to provide a standard architecture to be used by the whole manufacturing industry, but rather a reference architecture from which particular architectures can be derived which will fulfill the needs of particular enterprises.

The reference architecture provides constructs for the structured description of business requirements and for CIM system design and implementations. CIMOSA compliant enterprise systems will support organizational and operational flexibility, extensive use of multidisciplinary enterprise information (knowledge) and graceful system integration.

Through the business modelling framework a generic modelling concept is provided which is applicable to enterprises in many industries. This is a modelling concept which will vastly enhance communication and interoperability via Information Technology within an enterprise, as well as between different enterprises. Model execution in heterogeneous manufacturing and IT environments is supported by the implementation of an integrating infrastructure.

CIMOSA supports new paradigms in enterprise management, thus enabling the explicit description of enterprise processes at different levels of abstraction for use for strategic, tactical and operational decision support. Decision support is developed through the extensive simulation of alternatives and through the evaluation of decision impacts on all relevant parts of the enterprise. Specific definition of responsibilities and authority will allow for the deferring of decisions in time and for the reduction of the number of management levels in order to enable a flexible and efficient enterprise operation.

Applying CIMOSA modelling methodology results in complete descriptions of enterprise domains and of their contained business processes including their relationships to external agencies (suppliers, customers, government regulatory bodies, etc). This enterprise model is stored on and manipulated by the relevant information technology base of the enterprise. CIMOSA allows the modelling of the enterprise to be done incrementally rather than by following an overall top-down approach. It structures the enterprise operations into a set of interoperating domain processes exchanging results and requests. Several different views of the manufacturing enterprise content and structure are required to satisfy the needs of the different users of such an architecture. CIMOSA provides the necessary constructs to enable these multiple views to be created and manipulated by those users who have specialist knowledge of their particular field but who are not experts in IT.

6.11 SYSTEM LIFE-CYCLE

Enterprises are very complex undertakings which can be rendered more manageable by structuring their complex operational domains into a set of interoperating modules and its time domain according to a life-cycle concept. CIMOSA employs the concept of domain and business processes for its modular structuring and those of system and product life-cycle for the time domain.

Product life-cycle is a well known notion which is used to structure the different phases of a product's life. Within a given manufacturing enterprise many such product life-cycles exist – one for each product – which gives rise to a set of overlapping life-cycles as shown in Fig. 6.1. In order that the tasks associated with each step of the product life-cycle may be executed in a controllable manner, enterprises define the procedures to be followed. These normally take the form of company standards, work instructions, process sheets, etc. Different parts of the organization may be responsible for different phases. For example 'use' and 'maintain' is normally a 'client' rather than a 'supplier' task.

This principle of the life-cycle can be extended to the enterprise system or parts thereof (domains). While today's business processes are in operation, their successors may be in the process of implementation. Strategic

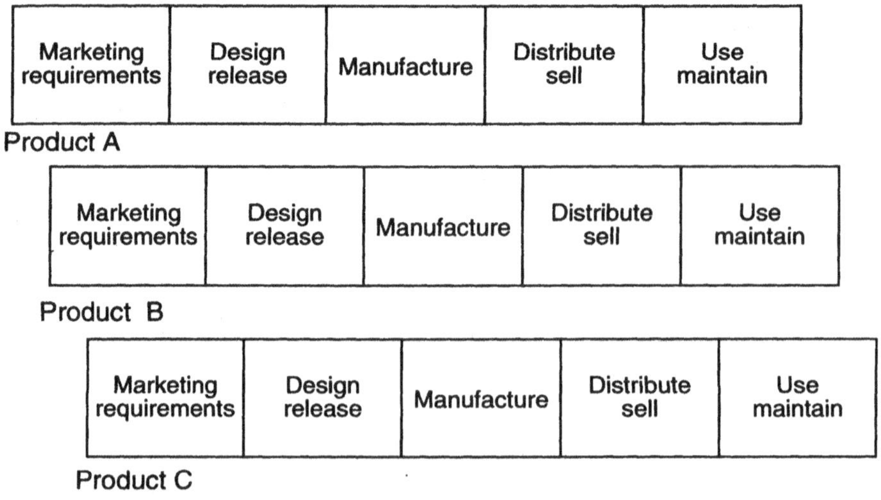

Figure 6.1 Overlapping product life-cycles.

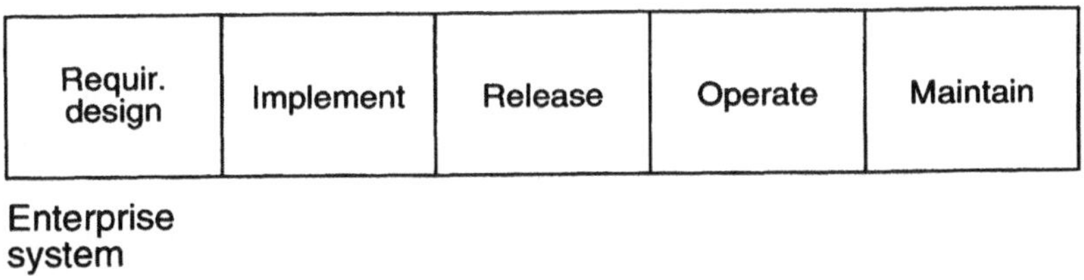

Figure 6.2 System life-cycle.

business plans under development today will set the stage for the systems implementations of tomorrow. Thus in an analogous way to products, the enterprise system which creates those products goes through a system life-cycle of development and use (operation) as shown in Fig. 6.2.

The tasks of the enterprise system life-cycle are concerned with the definition, description, creation and updating of the procedures and system components which govern and support the tasks of the product life-cycle. Thus product and system life-cycles are intimately interrelated. This relationship between the product and system life-cycles is summarized in Fig. 6.3. Whenever it is required to define a new or revised process supporting some part of a product life-cycle it will be necessary to carry out the tasks belonging to one or more phases of the associated system life-cycle. The operational phase of the system life-cycle is the execution of the relevant product life-cycle phases themselves. Thus by assigning enterprise tasks to one or the other of the life-cycles it is possible to structure the workings of the enterprise into enterprise engineering and operation and so render more effective the management of change.

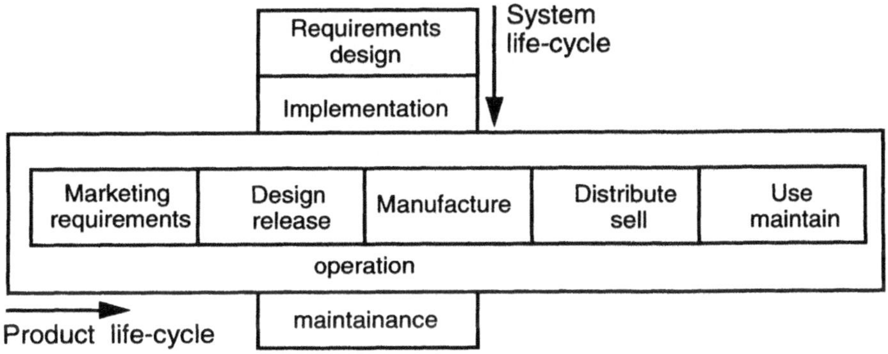

Figure 6.3 Relationship between system and product life-cycle.

6.12 ENTERPRISE INTEGRATION

From the foregoing it will be recalled that enterprise integration has to be an ongoing process rather than a one time effort. The level of integration should remain a managerial decision and should be open to change over a period of time. Hence, one could find in some parts of a CIM enterprise a set of tightly coupled systems and elsewhere a set of loosely coupled systems according to the choices made by this particular enterprise.

The need to implement multivendor systems both in terms of hardware and software and of easy reconfigurability requires the provision of standard interfaces. To solve the many problems of the industry, integration has to recognize this and proceed on more than one operational aspect. The AMICE project identifies, and addresses, three levels of integration covering the aspects of physical systems, of application and of business integration (see Fig. 6.4).

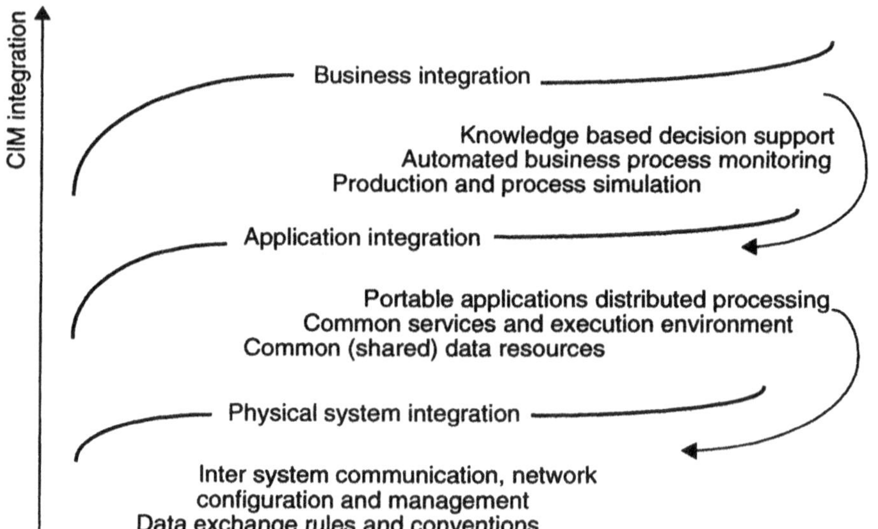

Figure 6.4 The levels of enterprise integration.

6.12.1 Business Integration

Business Integration is concerned with the integration of those functions which manage, control and monitor business processes. These are functions which provide for supervisory control of the operational processes and in turn coordinate the day-to-day execution of activities at the Application level. The modelling of business processes and of their interrelations and the use of the resulting models for decision and operational support is the key to business integration.

6.12.2 Application Integration

Application Integration is concerned with the control and integration of applications (applications in the dataprocessing sense). This means there has to be interoperation between applications and users (human beings as well as machines) and the supply and removal of information through inter and intra system communication. Integration at this level means providing a sufficient Information Technology infrastructure to permit a system wide access to all relevant information regardless of where the data reside.

6.12.3 Physical System Integration

Physical System Integration is concerned with the interconnection of manufacturing automation and data processing facilities (that is, join CAD to CAM, provide for flexible manufacturing cells, implement computerized scheduling, etc.) to permit the interchange of information between the so called 'islands of automation' (inter system communications).

Dictated by the needs of the individual enterprise this interconnection of physical systems was the first integration requirement to be recognized and fulfilled. This need has led to a number of information technology concepts becoming well accepted standards in this area[1].

---

1. For example: OSI (Open System Interconnection); MAP (Manufacturing Automation Protocol – an OSI implementation for the manufacturing shop floor).

6.13 CIMOSA OVERVIEW

To satisfy the above issues of Management of Change, of Flexibility and of Enterprise Integration, CIMOSA provides three interrelated concepts:

1. Modelling Framework (Reference Architecture, Particular Architecture, Enterprise Model).
2. System life-cycle and Environments (Engineering and Operation).
3. Integrating Infrastructure.

CIMOSA recognizes previous efforts in enterprise integration especially in the manufacturing industry and draws from the experience already gained in enterprise modelling and computer systems integration. Some of the efforts which have been of particular importance for the development of enterprise integration are given in Reference [8].

6.14 CIMOSA MODELLING FRAMEWORK

The CIMOSA Modelling Framework provides the necessary guidance to enable end users to model the enterprise and its associated CIM system in a coherent way. The CIMOSA modelling approach is based on a reference architecture from which Particular Architectures and Enterprise Models can be developed. The structuring and decoupling of user concerns from the implementation constraints provided by the framework contributes to enterprise flexibility.

The Modelling Framework provide a structure which clarifies the relationships between the parts which make up the enterprise operational system (Information Technology and Manufacturing Technology Components) and the methods and software tools which are required to describe, simulate and operate such an industrial system.

When modelling an enterprise there are many aspects and viewpoints to be examined that cannot be structured in a one dimensional framework. CIMOSA identifies a three dimensional framework offering the ability to model different aspects and views of an enterprise (see Fig. 6.5):

1. The genericity dimension (Building Blocks) is concerned with the degree of particularisation. It goes from generic building blocks to their aggregation into a model of a specific enterprise domain. This dimension differentiates between the Reference and Particular Architectures.
2. The modelling dimension provides the modelling support for the system life-cycle starting from statements of requirements to a description of the system implementation.

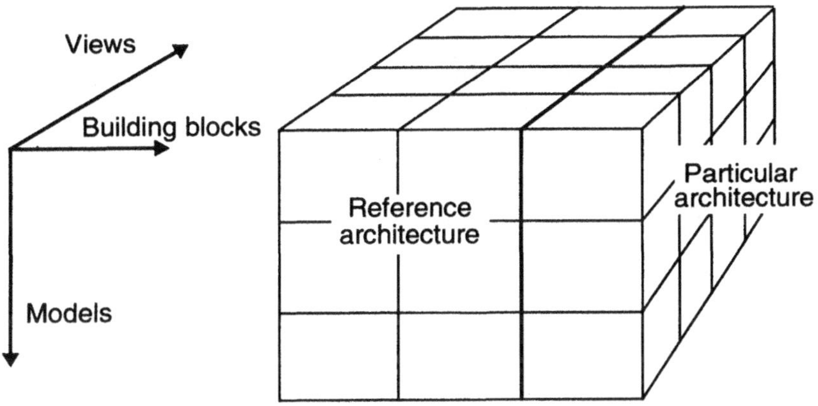

Figure 6.5 The CIMOSA modelling approach.

3. The view dimension is concerned with system behavior and functionality. This dimension allows the user to work with submodels representing different aspects of the enterprise (function, information, resource, organization).

6.14.1 CIMOSA reference architecture

Within the Modelling Framework, CIMOSA provides the CIMOSA Reference architecture. It resembles a catalogue of reusable building blocks which contains generic building blocks and aggregations of generic building blocks (Partial Models) applicable to specific modelling needs. See Fig. 6.6. User Oriented Generic Building Blocks are provided for:

Behaviour	Domain and Business Process, Event, Procedural Rules, Objective, Constraint
Function	Enterprise Activity, Functional Operation (input/output for function, control, resource), Objective, Constraint
Information	Enterprise Object, Object View, Information Element, Resources:
Resources	Capability, Functional Entity, Resource Unit, Cell

Organization:	Responsibility, Authority, Organizational Unit, Organizational Cell

IT Oriented Generic Building Blocks complement user building blocks supporting IT system design and implementation description. Such building blocks are Schemas, Transactions, Configurations and specific Functional Entity types representing the Integrating Infrastructure.

Partial Models are aimed at particular types of enterprise domains or even types of enterprises. Such partial models are created using the set of generic building blocks. Such partial models will be provided by vendors to be used as macros to enhance the model building process and as implementable components for system implementation.

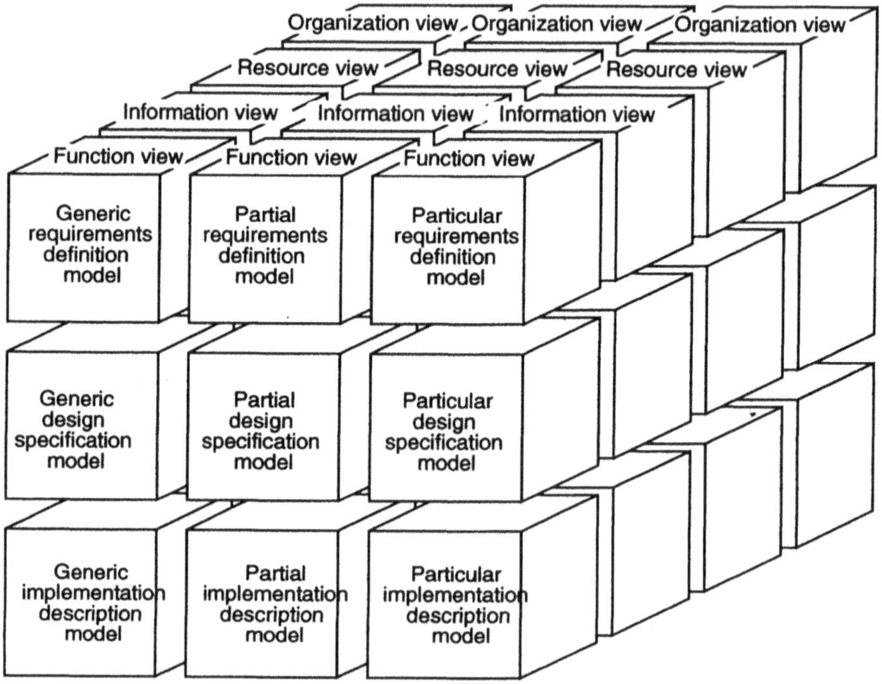

Figure 6.6 Overview of the CIMOSA architectural framework.

6.14.2 CIMOSA particular architecture

The structuring of particular enterprise domains follows the framework of modeling levels and views. In the Process/Activity/Operation approach an enterprise is seen as a collection of interrelated but nonoverlapping domains, which are subsets of the enterprise business. Enterprise engineering is concerned with requirements, design and implementation of enterprise operations.

Enterprise operations are defined in two distinct levels linked to each other by events:

1. A network of event driven cooperating domain processes
2. An explicit description of the functionality and behavior of the individual domain processes presented as a network of enterprise activities/functional operations initiated by events and driven by an explicit set of procedural rules.

The network of domain processes may consist of CIMOSA and non-CIMOSA domain processes, that is, domain processes with or without an internal CIMOSA structure. This approach provides a potential solution for the enterprise integration of heritage systems. Figures 6.7 and 6.8 represent the two levels of such structuring. Fig. 6.7 shows a set of domain processes cooperating on shared objects (results and events). One of those processes is detailed in the following figure demonstrating the engineering process of decomposition (top-down) leading to a network of enterprise activities. It also shows the use of function aggregation (bottom-up) into Business and domain processes. Decomposition can be either into business processes representing lower level behavior or enterprise activities representing functionality. For a system design enterprise activities can be further decomposed into functional operations which relate this functionality to enterprise resources for execution.

6.14.3 CIMOSA enterprise model

According to the structure provided for the particular architecture, CIMOSA models capture business knowledge globally as networks of domain processes and also as enterprise activities representing detailed local functionality and as business processes representing intra process behavior. Identification of all the inputs required and the outputs produced (information, control, resource and organizational) will be done for each of the different parts of this functionality. Modelling is done through the instantiation of generic building blocks or partial models.

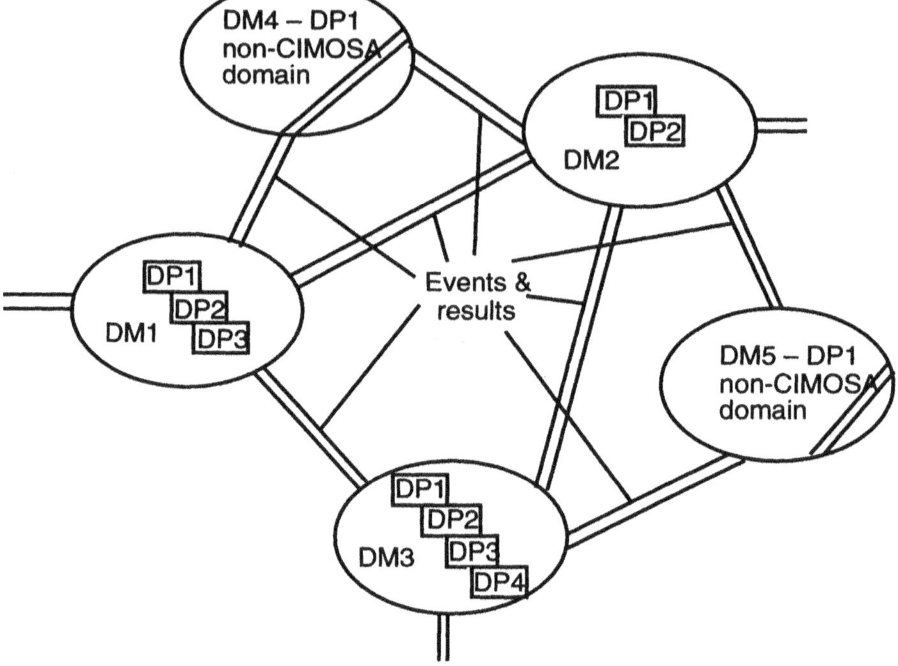

Figure 6.7 Network of domain processes (DM = domain, DP = domain process).

Within an enterprise, tasks (domain processes) are usually organized into subtasks (business processes, enterprise activities, functional operations) which need to be realized to achieve the business objectives. Within each domain (unsolicited) real world happenings and requests (events) to do something will occur which will trigger domain processes. Domain processes can also have exchange with domains external to the enterprise. To represent tasks and actions performed within an enterprise CIMOSA forces users to think in terms of 'processes', 'activities' and 'operations', where operations define the lowest level of granularity to represent tasks performed within an enterprise. The level of detail to be described in the model is at the users discretion and not dictated by CIMOSA.

Model modifications can be made via a domain process' internal and external behavior and its functionality. Internal changes will be made through modification of internal control flow and/or functionality, changes in external behavior will be modelled through modifications of its external relations (interoperation with other domain processes).

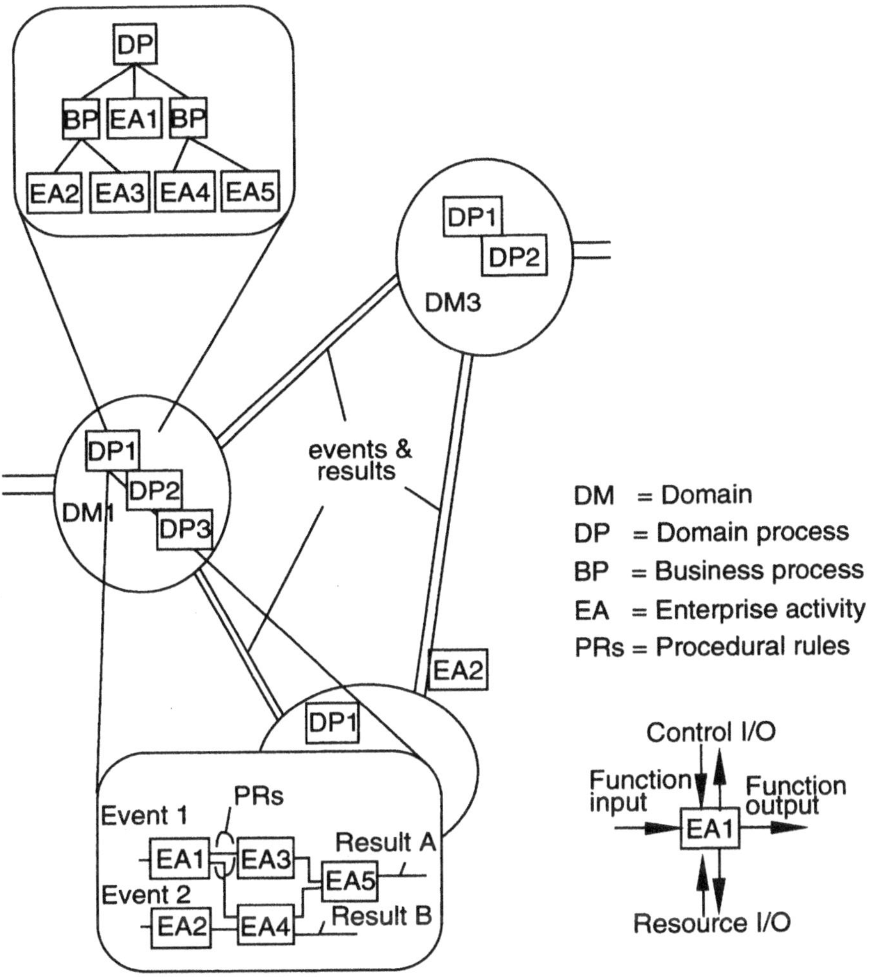

Figure 6.8 Internal structure of a domain process.

The CIMOSA modelling approach includes and extends previous functional modelling approaches [9] already in use in industry or being considered by standardization bodies. These recognize essentially only the concept of activity (also called function).

CIMOSA differentiates between 'as-is' and 'to-be' modelling. Modelling of an existing implementation will start with a bottom up description of the current operation. Abstraction from the 'as-is' description and modifying the operation would lead to a requirements and design of the 'to-be' model and analysis of its intended behavior. No specific methodology has been prescribed by CIMOSA thus leaving freedom for iterations as required between decomposition and aggregation as well as between modelling levels.

6.15 CIMOSA SYSTEM LIFE-CYCLE AND ENVIRONMENTS

To support its modelling concept, CIMOSA defines an enterprise system life-cycle and the corresponding environments in which the relevant life-cycle processes can be executed. These environments contain: (1) a dynamic definition (a model) of the life-cycle tasks in the form of a set of processes; (2) a set of resources providing the functionality required to carry out those tasks and (3) a set of common services providing system wide support for the execution of processes. This concept decouples the engineering of the enterprise's system(s) from the day-to-day enterprise operations.

In order to permit the processes for one task within the Product Life-cycle to be designed/updated without interfering with the execution of current operational tasks, CIMOSA provides two mutually independent, but interrelated, environments (see Table 6.1).

Table 6.1 CIMOSA life-cycle and environments

Life-cycle	*Environment*	*Model executed*
system	enterprise engineering	engineering implementation model
product	enterprise operation	particular implementation model

1. The Enterprise Operational Environment in which products are developed, produced, marketed and sold, orders are issued and received, bills are paid and invoices submitted to customers in accordance with the processes defined for the enterprise's Product Life-cycles.
2. The Enterprise Engineering Environment in which particular enterprise domains are modelled (including the simulation/evaluation of alternative solutions), built, validated and released for operation in accordance with the processes defined for the enterprise's system life-cycle (except for the system operation phase).

In both cases life-cycle processes are defined by a CIMOSA compliant model. In the case of the enterprise operations environment this model is user created so as to be unique to the particular enterprise. The model of the enterprise engineering environment is usually constructed by the IT vendor and then customized to the user's needs. Conformance to the CIMOSA Model Creation processes and the CIMOSA Model Consistency rules ensure that the end result is a CIMOSA compliant model (see also Table 6.2).

Table 6.2 Service entities of the Integrating Infrastructures service provided

Entity	*Execution of enterprise model*
Business	According to business rules
Presentation	Homogeneous access to humans, machines, software applications
Information	High level of consistent information
Common	System wide homogeneous data communication
System management	Management of CIMOSA environments

CIMOSA defines the basic phases of the System Life-cycle in terms which are independent of the modelling methodology employed. CIMOSA does not populate the System Life-cycle with a set of unique processes but leaves it open for tool vendors to define their own various proprietary methodologies in a CIMOSA compliant model. Thus alternative models (tool sets) will exist for the system life-cycle to reflect different modelling methodologies (e.g. waterfall, fast prototyping, concurrent engineering, etc.) for creating CIMOSA conformant enterprise models.

For example, the creation and execution of a CIMOSA compliant process involves the following basic tasks:

1. Identify the business domain
2. Select relevant building block types from a catalogue of CIMOSA building blocks

3. Customize the selected types by adding enterprise specific parameters to make instances of the building block types
4. If required install and describe any additional resources
5. Release the enterprise model which contains the instances, into the operational system
6. Add, at execution time, technical parameters and logistic variables to a copy of the released instances (create occurrences).

CIMOSA proposes a formal engineering release of the enterprise model at the phase-over from the engineering to the operational environment. Changes to be made to the operational system have to be implemented, validated and documented in the engineering mode rather than in the operational mode. Only then can a sufficient engineering discipline be introduced into the enterprise operation and only then can model consistency be assured. This release process acts as a 'bridge' between the system and the Product Life-cycle.

It should be noted that emphasis in CIMOSA Enterprise Engineering is on model maintenance (update, modification, extension) rather than on model creation. The latter is seen as a one time affair, whereas the former is a continuous task for redefining the model contents. Only if the model continues to reflect the reality of the enterprise operation will the use of model based operation control and monitoring and decision support become a reality.

6.16 CIMOSA INTEGRATING INFRASTRUCTURE

CIMOSA copes with heterogeneity by providing the CIMOSA Integrating Infrastructure (IIS). This IIS provides a structured set of system wide services thereby avoiding redundancy of function in the enterprise's systems. By making use of these generic services the application systems in the CIMOSA Environments need no longer contain the specifics of the data processing environment. This provides increased portability and flexibility of the application systems and reduces the maintenance tasks considerably.

The IIS supports integration by providing homogeneous access to system components. The services of the CIMOSA Integrating Infrastructure address the problems of Application Integration. The IIS supports model execution and engineering through a set of five generic service entities (see Fig. 6.10). It also provides a unifying software platform to achieve integration of heterogeneous hardware and software components.

The CIMOSA Integrating Infrastructure is tuned to the execution of CIMOSA models. Triggering of domain processes invokes the Business

Entity which will process the request and which will initiate the creation of a domain process occurrence and its subsequent processing by other services according to the nature of the event. The services of the Business Entity will schedule resources and execute functional operations through the Presentation Entity. It will also provide and remove information via the Information Entity. Communication across the entire system and the needs of system management are provided by the remaining two entities (Common and System Management Entity).

The AMICE project defines functionality of the IIS services but its implementation is under the discretion of the IT vendors. The IIS may be implemented according to any standard or proprietary solution which provides the expected functionality. Access to the different services is specified by interfaces or protocols. Therefore, depending on the degree of software integration, the IIS software may be a very small extension or a large software packet added to the vendor system software.

The IIS provides a structured set of common System Wide Services that will avoid redundancy of functions in any CIM system. By offering these services system wide in a uniform manner the basis for required integration is created. The role of the IIS is to provide the computer integrated part of CIM, meaning that the integration is achieved through Information Technology, i.e. computers.

The IIS provides for a set of specified services to control and integrate the definitions in the implementation model of a CIM architecture and the resources (humans, machines, applications, etc.) performing these functions. The services will:

1. Isolate the application programs from the data processing and business environment
2. Operate system-wide and not require the architectural modules using these services to have any knowledge of their location and/or distribution
3. Provide common protocols allowing the humans, machines and application programs to interact with the IIS in a uniform way.

The System Wide Services appear to the user of the service as a single service across all nodes of the system. The user doesn't have to know how and where the service is provided. Within the services of the IIS one can distinguish three sets of services.

Function related services include all those services that are related to enterprise functions. These services manage the control and execute the enterprise activities. They provide the Integration of Functions.

Information related services include services that support the information processing activities of the enterprise. They deal with locating, accessing,

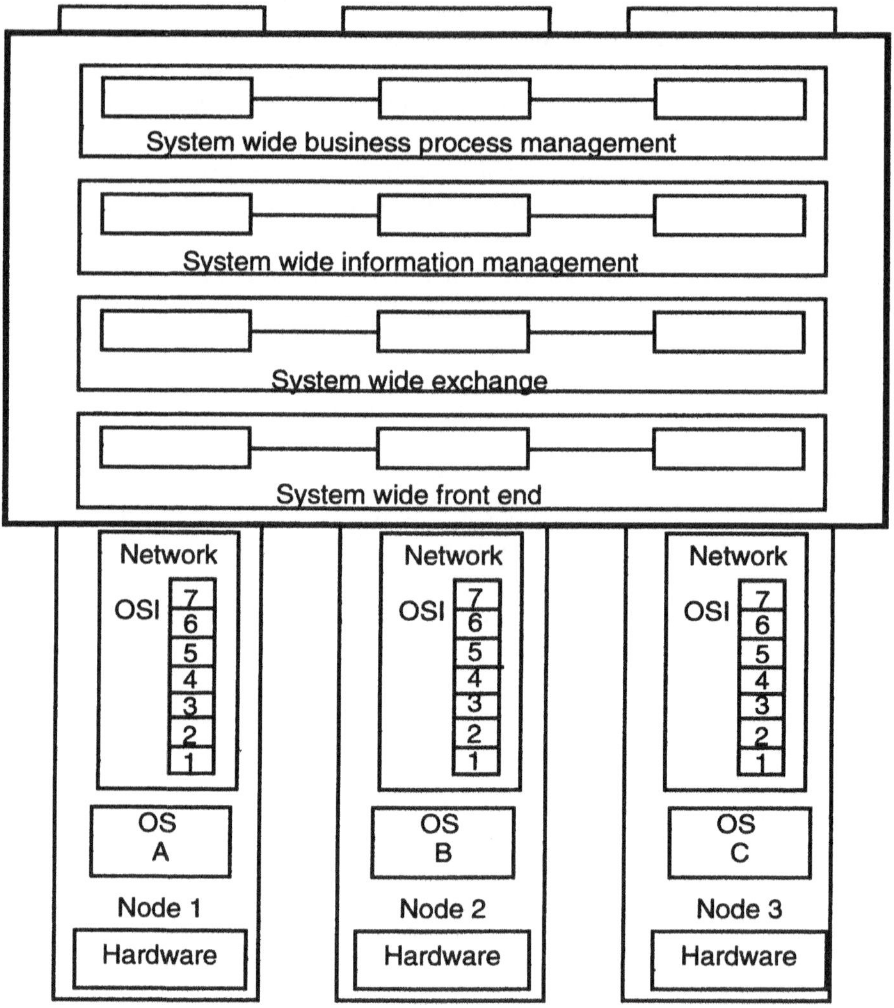

Figure 6.9 Overview of all the services within the IIS.

storing and maintaining the consistency of information. These services provide the Integration of Information.

Communication related services include the services required for the control of Intra- and Inter-System Communication. They provide the Integration of Communication.

Fig. 6.9 provides an overview of the IIS. The three sets of services that were first mentioned will be detailed hereafter.

The Function Related Services provide a common functional control service that integrates all functions of the enterprise along business processes oriented towards the result of the enterprise. The system wide business process management services (see Fig. 6.9) uses both resource and organizational views of the architectural model and contains human entities, machine entities, and application entities which are interfaced to the architecture via System Wide Front End Services through front-end protocols (see Fig.6.10).

Three Management Services have been defined for the proper functioning of the enterprise:

1. The business process control services which manage and control the business processes of the enterprise on a system wide basis.
2. The Activity Control services which manage and control the execution of the activities in the enterprise on a system wide basis.
3. Resource Management services which use the resource information of the implemented model and occurrences of the business processes to schedule the resources and the business processes on a system wide basis.

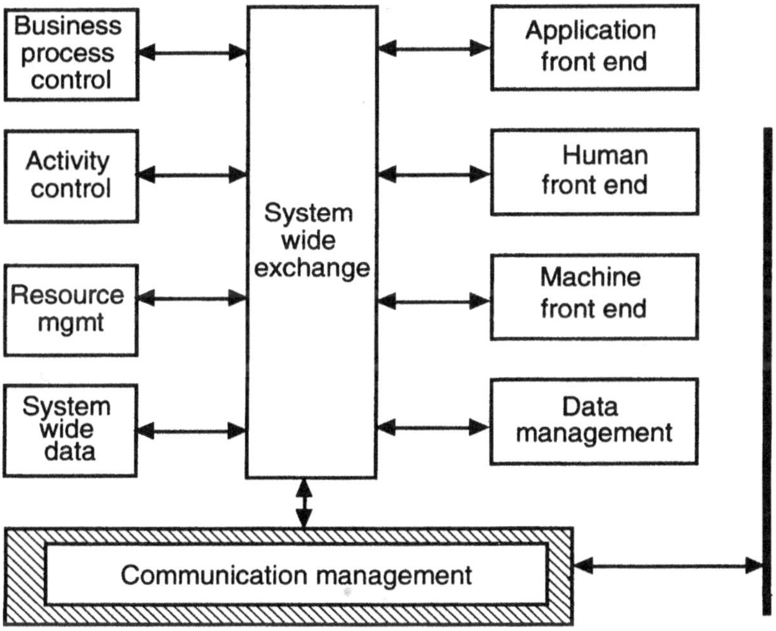

Figure 6.10 Services of the integrating infrastructure.

The purpose of the Information Related Services is to manage the storage and retrieval of data on a system wide basis where the clients don't need to know how and where the data is stored. The services further provide system wide data consistency and information context.

The System Wide Data Service provides for the location of information, consistency of data, management of schema conversions and control of requests for data. It also provides for storage and retrieval of data through Data Management. The System Wide Data Services handle all requests for information from clients. The Data Management has only one client; the System Wide Data Services. Data Management interacts with the supporting functional entities (business entities, information entities, presentation entities and human and machine entities) for the storage and retrieval of data and caters for the further completion of the schema conversions.

The purpose of the Communication Related Services is to allow all implemented functional entities to communicate with each other on a system wide basis. These entities do not have to have any knowledge of where the other entity is located and how the locations communicate with each other. There is no difference between inter- and intra-system communication for the user of this service.

System Wide Exchange Services will provide an adequate service needed to implement access, agent and external protocols occurring among information related services and function related services, while factoring out as many as possible of the issues of synchronous communication.

Communication management, finally, has one client only: the System Wide Exchange Services. It interacts with the supporting implemented entities to send and receive data based on the concepts developed in the Basic Reference Model for Open System Interconnect (ISO 7498).

6.17 APPLYING CIMOSA – A SUMMARY

Fig. 6.11 illustrates the use of CIMOSA. The reference architecture provides the constructs (Generic Building Blocks) and standardized processes (Partial Models) which are used to create and maintain the particular enterprise model and to implement the IIS. The enterprise model describes the Product Life-cycle processes (domain processes) of the particular enterprise and their interrelations. These processes are composed of enterprise activities and functional operations. Resources are selected from vendor catalogues or from the installed enterprise base.

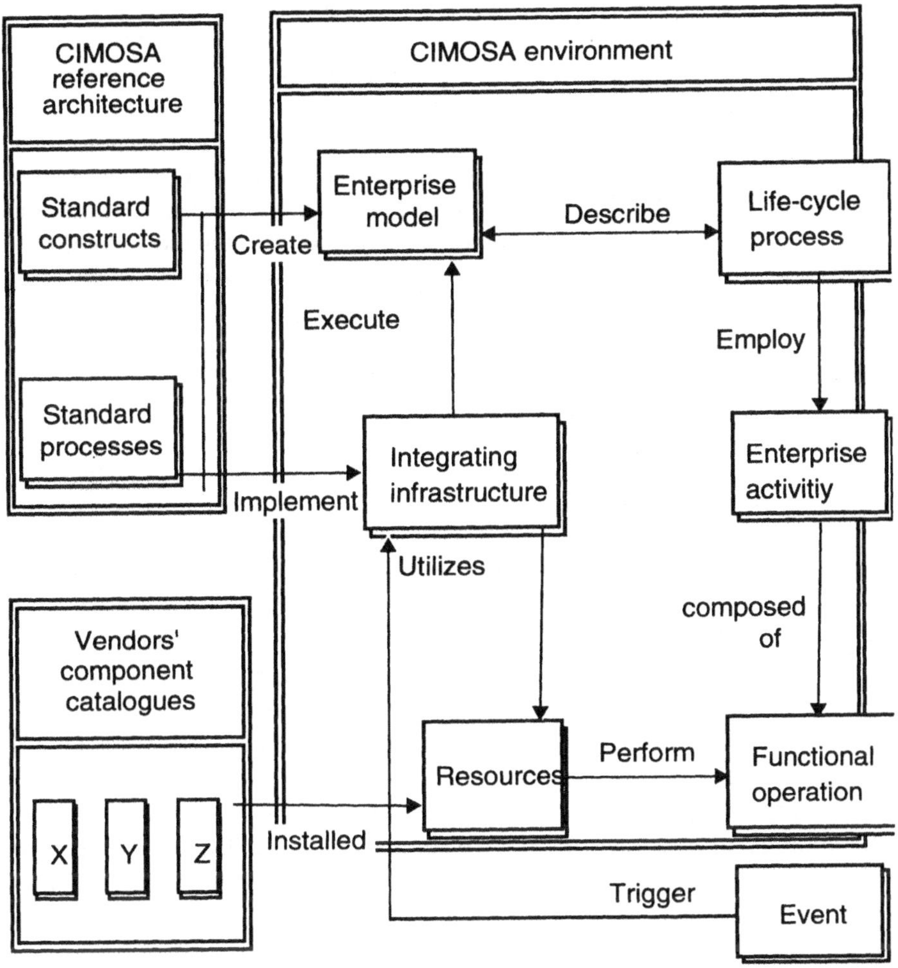

Figure 6.11 Creation and execution of CIMOSA models.

Model creation, implementation and maintenance are part of the engineering environment and follow the System Life-cycle concept. Model execution is concerned with the processes of type Product Life-cycle. These processes are performed through the utilization of relevant resources which are identified in the different processes as resource inputs of the enterprise activities. Model execution is initiated by events which represent real world happenings like receiving a customer order or identifying a due date. Any event will start execution of at least one domain process (see above).

Model based operational control and monitoring requires a consistent and up-to-date model of the operation. Therefore, modelling has to be done by the people responsible for the operation and authorized to modify and adapt the particular part of the model according to negotiated and accepted needs and opportunities. Knowledge about interrelations and the impact of change have to be made known to the parties involved in the engineering phase and prior to release to operation. Therefore analysis and evaluation of change is an important part of model creation and maintenance and has to be employed by all the people involved in enterprise engineering. Assignments of responsibility and authority in the engineering environment are very important and have to be made with a clear understanding of the consequences involved.

6.18 OVERALL SUMMARY OF ARCHITECTURE AND BASIC CONCEPTS OF CIMOSA

Today the manufacturing industry is faced with a rapidly changing business environment. Varying market demands require frequent operational and organizational changes in the individual enterprise. CIMOSA therefore enables the user to describe his enterprise from two distinct standpoints:

1. The Business Environment as seen by the business user
2. The Physical Environment (the manufacturing and Information Technology implemented in the enterprise).

The prime goal of CIM and enterprise integration is to use Information Technology to integrate the enterprise operations. CIMOSA achieves this by linking the enterprise activities and functional operations into a dynamic state network, thus integrating the activities and the resources performing them. Fig. 6.12 summarizes the contents and use of CIMOSA in its two environments and the relations of its three major concepts with one another and the real enterprise world.

1. Modelling Framework
2. System Life-cycle
3. Integrating Infrastructure

CIMOSA provides a reference architecture in its modelling framework for the specific description of a particular enterprise. The architectural constructs of CIMOSA guide the CIM business user in the creation and maintenance process to obtain and maintain a consistent system description. These constructs will also guide CIM suppliers to develop, produce and market products which will fit the defined user requirements and comply with the implemented system.

The architectural constructs provided for the description can be viewed as the semantics for a manufacturing enterprise description language. This approach constitutes a completely new method for manufacturing system development. The method allows the end user to define, to prototype, to design, to implement and to execute the business according to his needs. The architecture also defines the necessary support environment for development and execution.

CIMOSA facilitates a System Life-cycle which guides the user through model engineering and model execution. Starting with the collection of business requirements in a Requirements Definition Model, the System Life-cycle carries the user through their translation into a System Design Model and finally, to the description of the implemented system (System Implementation Model). These phases are followed by a model release for operations and model execution for operational control and monitoring. To derive the system description all specific constraints of the particular enterprise are taken into account in the Design Specification Model. Model maintenance is a very important part of the System Life-cycle as well.

Model engineering and model execution are supported by the CIMOSA Integrating Infrastructure. The Integrating Infrastructure provides a set of generic services which process the released enterprise model and provide access to information, connect to resources and handle exceptions through reference to identified responsibilities and authorities. This infrastructure also hides the heterogeneity of the underlying Manufacturing and Information Technology.

The ESPRIT Consortium AMICE has developed CIMOSA and has validated its concepts in various case studies in its member companies. Validation work has also been done in other ESPRIT projects which will also lead to development of partial models in specific industry areas and particular modelling aspects. The project itself will further verify CIMOSA through industrial pilot implementation. Work on prepilots in enterprise modelling and model based operation control and monitoring has been started. Results from this work will become available in 1993.

6.19 REFERENCES

[1] Anonymous, AMICE: CIMOSA, ESPRIT Project 5288, Milestone M-2, AD2.0 Volume 2, Architecture Description, Document RO443/1, Consortium AMICE, Brussels, Belgium, (August 24, 1992).

[2] H.T. Goranson. 1992. 'Dimensions of Enterpries Integration,' in [3] pp101–113.

[3] C. Petrie Jnr (Ed), 'Enterprise Integration Modeling,' Proc. First International Conference on Enterprise Integration Modeling, MIT Press, Cambridge, MA, USA (1992).

[4] CEN/CENELEC (The Joint European Standards Organization), European Prestandard Env 40003, 'Computer Integrated Manufacturing, System Architecture, Framework for Enterprise Modelling,' Berlin, Germany, (April 18, 1990).

[5] ESPRIT, Project 688, AMICE Open System Architecture for CIM, Springer Verlag, Berlin (1988).

[6] Publications by AMICE Project members

(a) Jorysz, H. R. and Vernadat, F. B., 'CIMOSA Part 1: Total Enterprise Modelling and Function View,' Int. J. Computer Integrated Manufacturing, Vol. 3, Nos. 3 and 4, pp. 144–156 (1990).

(b) Jorysz, H. R. and Vernadat, F. B., 'CIMOSA Part 2: Information View,' Int. J. Computer Integrated Manufacturing, Vol. 3 Nos. 3 and 4, pp. 157–167 (1990).

(c) Klittich, M., 'CIMOSA Part 3: CIMOSA Integrating Infrastructure – The Operational Basis for Integrated Manufacturing Systems,' Int. J. Computer Integrated Manufacturing, **3** (3–4) pp. 168–180 (1990).

[7] ESPRIT Consortium AMICE Open System Architecture, CIMOSA, AD 1.0, Architecture Description, ESPRIT Consortium AMICE, Brussels, Belgium (1991).

[8] A Set of General Works in Enterprise Integration Described in Numerous Publications.

(a) U.S. Air Force (no author) 'Integrated Computer-Aided Manufacturing (ICAM),' Architecture Part II, Vol. V – Information Modeling Manual (IDEF) AFWAL-TR-81-4023, Wright Patterson Air Force Base, OH, (June 1981). Plus other volumes in this same series.

(b) Throop, J. W. and Read, P. R., A Discrete Parts Manufacturing Model, Report Number R-83-SC-01, Computer Aided Manufacturing – International, Inc., Arlington, Texas (August 1983).

(c) Simpson, J. A., Hocken, R. J., and Albus, J. S., 'The Automated Manufacturing Research Facility of the National Bureau of Standards,' Journal of Manufacturing Systems, Vol. 1, No. 1, pp. 17–32 (1982).
(d) ISO (International Organization for Standardization), ISO TC184/SC5/WG1, The Ottawa Report on Reference Models for Manufacturing Standards, Report N51 (September 14, 1986).
(e) ESPRIT Pilot Project 34: Design Rules for Computer Integrated Manufacturing Systems.
(f) ANSA: Advanced Networked System Architecture.

[9] General works in enterprise modelling (numerous publications):
(a) Softech Inc., ICAM Architecture of Manufacturing, Integrated Computer-Aided Manufacturing (ICAM), Task I – Final Report, AFML-TR-78-148, Waltham, MA (November, 1978).
(b) Ross, D. T., 'Applications and Extensions of SADT,' IEEE Computer, Vol. 18, No. 4, pp25–34 (April 1985)
(c) Doumeingts, Guy, Méthode GRAI: Méthode de Conception des Systemes in Productique. These d'Etat Automatique, Universite de Bordeaux 1 (November 1984).
(d) Bravoco, R. R., and Yadav, S. B., 'Requirements Definition Architecture – An Overview',Computers in Industry, Vol. 6, No. 4, pp. 237–251 (August 1985).
(e) Bravoco, R. R. and Yadav, S. B., 'A Methodology to Model the Functional Structure of an Organization,' Computers in Industry, Vol. 6, No. 5, pp. 345–361 (October 1985).
(f) Ward, P. T., and Mellon, S. J., Structural Technologies for Real-Time Systems, Vol 1–5, Yourdon Press/Prentice–Hall, New York, NY (1985).

6.20 FURTHER READING

AMICE Consortium, Open System Architecture for CIM Research Reports of ESPRIT Project 688, Volume 1, Springer Verlag (1989).

Beekman, Dirk, 'CIMOSA: Computer Integrated Manufacturing – Open System Architecture, Int. J. Computer Integrated Manufacturing, Vol. 2, No. 2, pp. 94–105 (March–April 1989).

Canter, J. D., Vectoring Towards 2001, A Strategic Blueprint for the Competitive Enterprise,

CEN/CENELEC/WG ARC N150, Evaluation Report of Generic Constructs for Function Views and Other Views in ENV 40003, Brussels, Belgium (January 1992).

Charue F., Apprentissages Organisationnells et Mutation Industrielle. Thèse, Universite de Paris (May 1991).
Chen, D. and Doumeingts, G., A Comparative Study Between CIMOSA and GIM, Laboratorie GRAI, University of Bordeaux, Bordeaux, France (May 1991).
Gallois,P.M., Le Manegement Industriel, Abance, Paris (1991).
Anonymous, Made in America, MIT Press, Cambridge, MA, (1989).
Roboam, M., La Méthode GRAI, Aerospatiale Paris, (1991).
Simon, H. A., Administration et Processus de Decision; Gestion Economica Paris, Paris, France (1983).
Tourratier, J. M., Les Résultats de l'Intégration Phase par Phase, ADEPA, Pari(1991).
Ullman, J. D., Principles of Database Systems, Second Edition, Computer Science Press, Rockville, MD (1982).

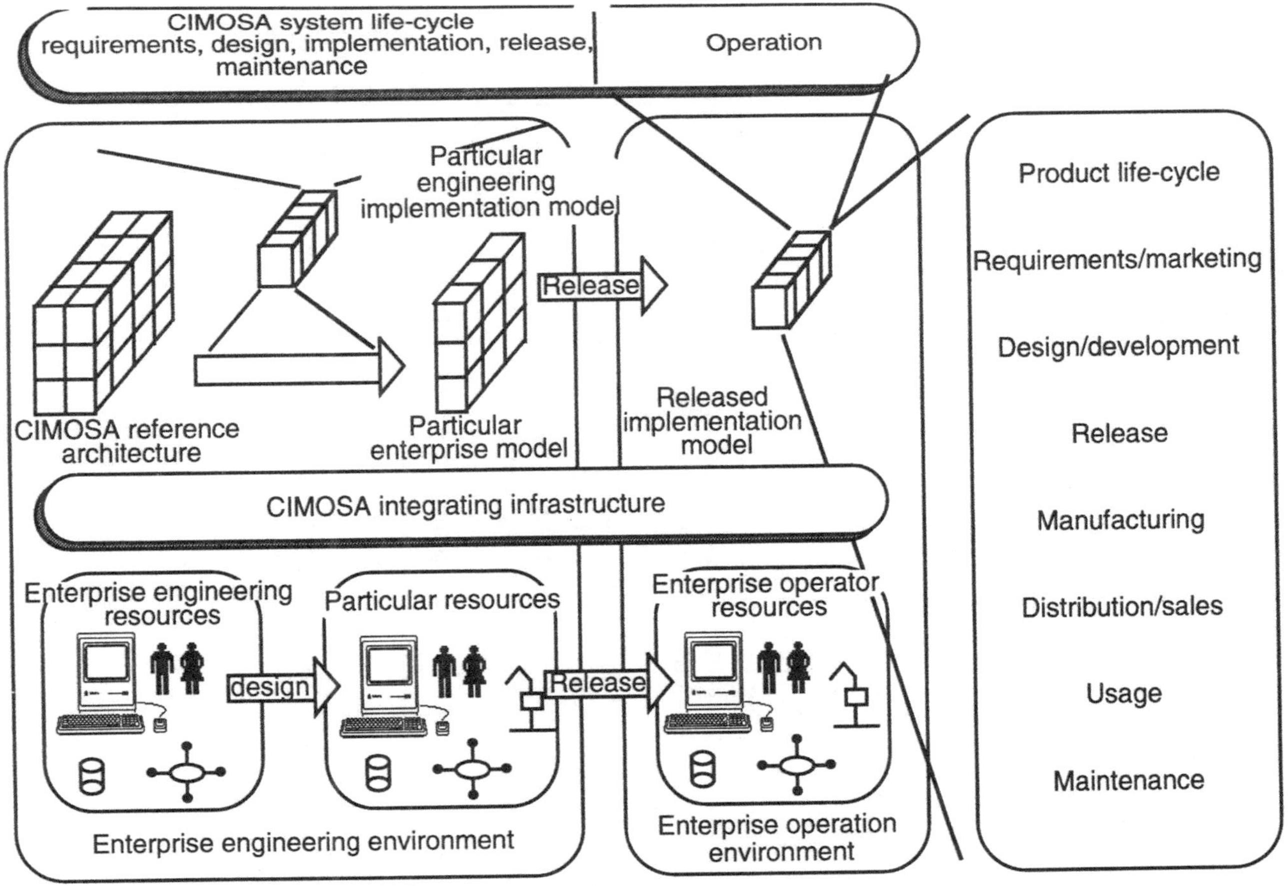

Figure 6.12 Overview of CIMOSA concepts.

7

The GRAI-GIM reference model, architecture and methodology

D. Chen and G. Doumeingts

The GRAI models [1–3] give a generic description of a manufacturing system while focusing on the details of the control part of this system. The control of a manufacturing system will be presented first from a global point of view and afterwards at the level of the decision center.

In order for this description to be generic enough to be usable for any type of manufacturing system, the theoretical part is very important.

The work on the GRAI-GIM methodology started in the 1970s at the GRAI Laboratory of the University of Bordeaux. The objectives at that time were to model a production management system in order to be able to define precisely the specifications needed to chose a software package for a Computer Aided Production Management (CAPM) system.

At that time GRAI already worked closely with industry and we were surprised to learn of the difficulties involved in introducing CAPM packages in Industry. Our first conclusion was that people gave too much attention to the term C.A. (Computer Aided) and less to the term P.M. (Production Management). Coming from the field of Automatic Control we knew that the first task before us in introducing Automation was that of the modelling of the system concerned.

The original GRAI Model was first developed to help the designer to model a Production Management System. With the subsequent development of CIM systems, we have extended the same GRAI model to the whole Manufacturing System and have extended the GRAI Method to the design of such a CIM system. This became GIM (GRAI Integrated Methodology).

The domain of GRAI-GIM is to support the designer of a CIM system to help him elaborate the model of an Integrated Manufacturing System in order to deduce the specifications of that CIM system. These components will either be bought or developed in house.

The need for a methodology to design a CIM system is clearly based on the concept of a CIM system as presented below: A CIM system

- Can not be bought
- Must be build for each enterprise
- Must be developed based on existing and reliable components.

The development of GRAI-GIM was carried out in close cooperation with Industry.

The first application of the GRAI method (for Production Management) was made in 1980 with Telemecanique Electrique. Between 1980 and 1985 thirty other applications of the GRAI method were realized either by the GRAI Laboratory itself or by other Engineering firms.

In 1985, SNECMA, one of the most important manufacturers of Aircraft Engines used the GRAI method to design and specify the production management system of a Flexible Factory in Le Creusot, France.

In 1986, the GRAI Laboratory started to work on ESPRIT (European Strategic Programme for Research and Development in Information Technology) projects. Further development of the GRAI method has been carried out within this framework. The results obtained after several projects (Open CAM System (1986), IMPACS (1989), FOF (1989), Flexquar (1992)) have been the extension of the GRAI method to GIM (GRAI Integrated Methodology).

Today, there also exists the organization called AUGRAI (Association of the Users of GRAI Method) with about twenty industrial members (SNECMA, Aerospatiale, Cap Gemini, Pechiney, etc) which supports the further development and the promotion of GRAI-GIM.

If we compare the GRAI model to other architectures we find some similarity with the CIMOSA architecture concerning their concepts of views. The presentation of GRAI-GIM contains three parts:

- Conceptual reference model called GRAI
- Formalisms to describe the GRAI model
- Structured approach.

7.1 THE GLOBAL MODEL

From the considerations expressed above, we can say that a manufacturing system may be split into two parts (Fig. 7.1):

- On the one hand the physical system composed of the people, facilities, materials and techniques which has the aim of transforming the raw material components into a final product in order to add value to the material flow:

- On the other hand the control system whose purpose is to control the physical system in accordance with the objectives defined.

The control system can also be split into two subsystems (Fig. 7.2)

- A decision system
- An information system.

The goal of the first one is to make the necessary decisions in order to define the orders to be transmitted to the physical system; the second one enables the transmission, processing and memorization of the information needed. It acts as a link between the physical and the decision systems and with the environment as well.

The decision system and the information system are strongly linked. Thus a close synchronization between them goes far to assure the efficiency of the control system.

When a system has a high level of complexity it can be efficiently controlled by the implementation of a hierarchical control system. The management theory of such a system introduces three typical levels which are: strategic, tactical and operational ones (Fig. 7.3). In such a system each level must be coordinated by the next upper level.

Another aspect we have to take into account is that coming from classical management theory which emphasizes the functional decomposition of a such system. This approach presents several functions (the number being related to the complexity of the company) which are closely linked to the organization of the company according to services being performed. This

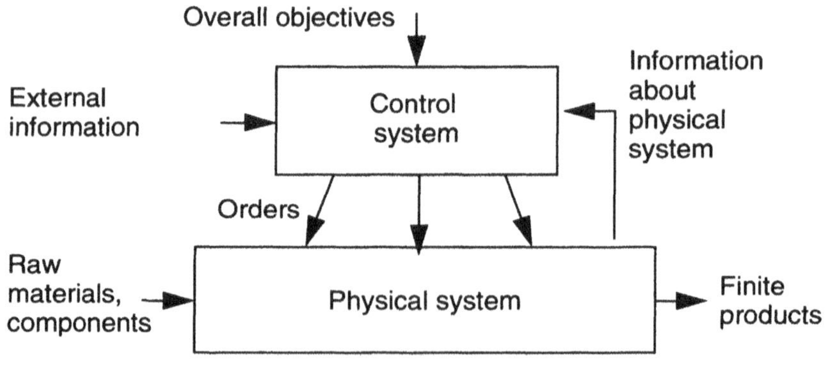

Figure 7.1 Splitting of a manufacturing system.

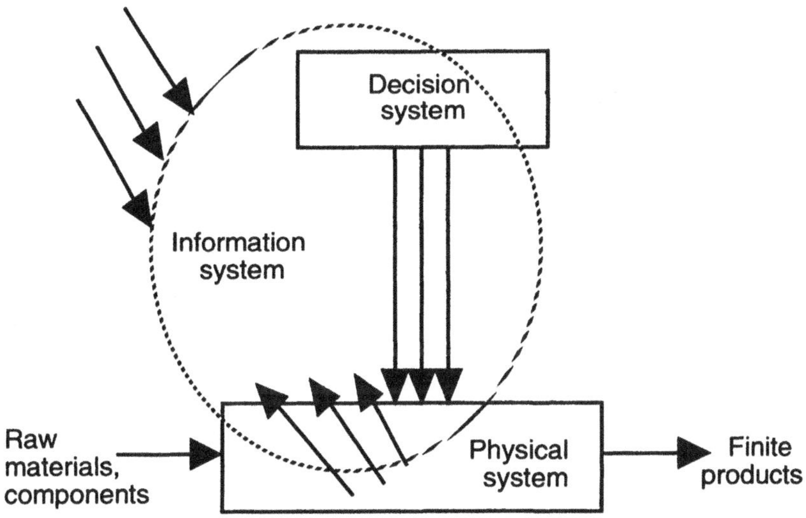

Figure 7.2 The decision and information systems.

approach then brings us a decomposition into three main functions: the resources management, the product management and the planning.

The first one is related to the transforming entities (the resources) which are made available by the structure of the system. The second one is linked to the major purpose of the factory system which is the transformation of products coming from the environment into new products for the environment. The planning function aims at coordinating all of the system by taking into account the global objectives of the company and making compromises between the workings of the other two functions as necessary (Fig.7.4).

The concept of feedback must be put into this model. Controlling the physical system without feedback means that the control system does not know the current status of the resources. Because the control system is hierarchial, all information from the physical system must move to each control level through the filter and aggregation functions of the level below (Fig.7.5).

When defining the control system of any manufacturing system these considerations lead us to propose a two dimensional decomposition:

1. A functional decomposition (vertical divisions), defining the various functions of such a system for each major part.
2. A hierarchical decomposition (horizontal divisions) in accordance with the time criteria of instituting decision making levels which are

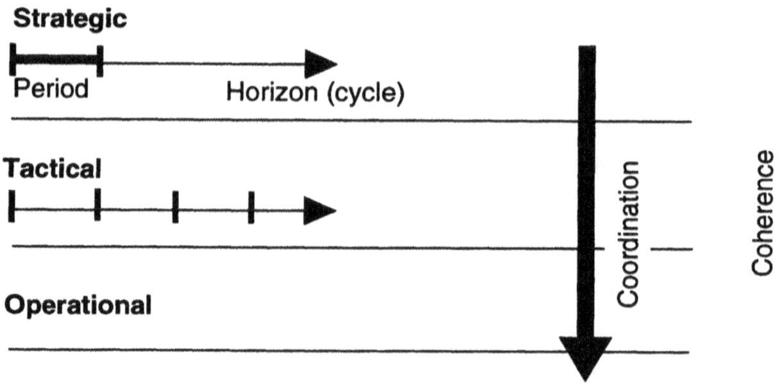

Figure 7.3 Hierarchical aspects: strategic, tactical and operational levels.

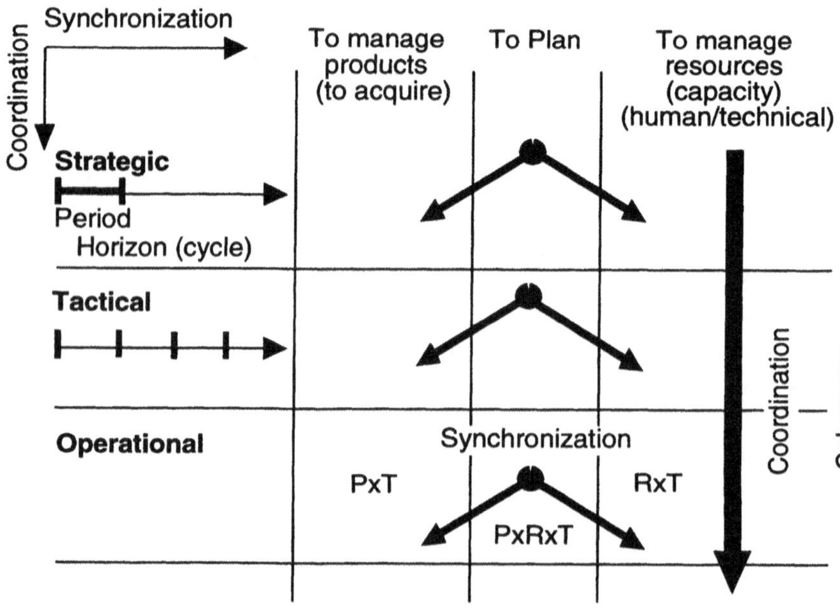

Figure 7.4 Hierarchical and functional aspects.

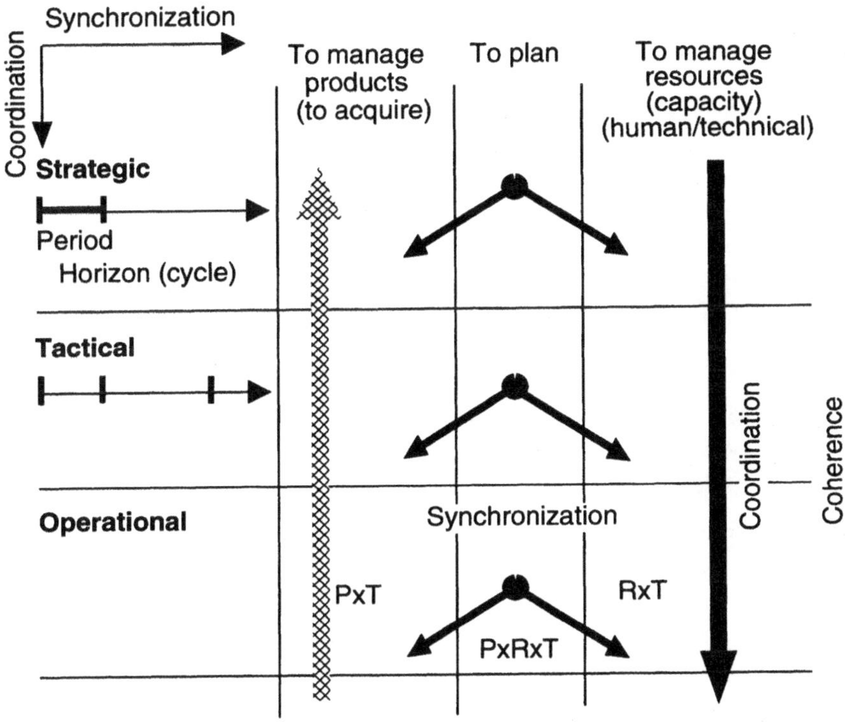

Figure 7.5 Feedback introduction.

characterized by the concept of a time horizon (the time interval during which the result of the decision is relevant) and the concept of a time period (the time interval after which the result of the decision must be reconsidered). This notion is a generalization applied to each of the three classical levels.

This two dimensional decomposition enables the concepts of a decision center to be defined. A decision center contains as a generality all of the decisions made within one function and at one hierarchical level. Fig. 7.6 shows these concepts globally.

The hierarchically organization of the decision system into decision centres which are inter-related makes the information system hierarchically structured as well. Internal information from the physical system and external information from the surrounding environment is filtered and aggregated as required by the hierarchical level having to use it.

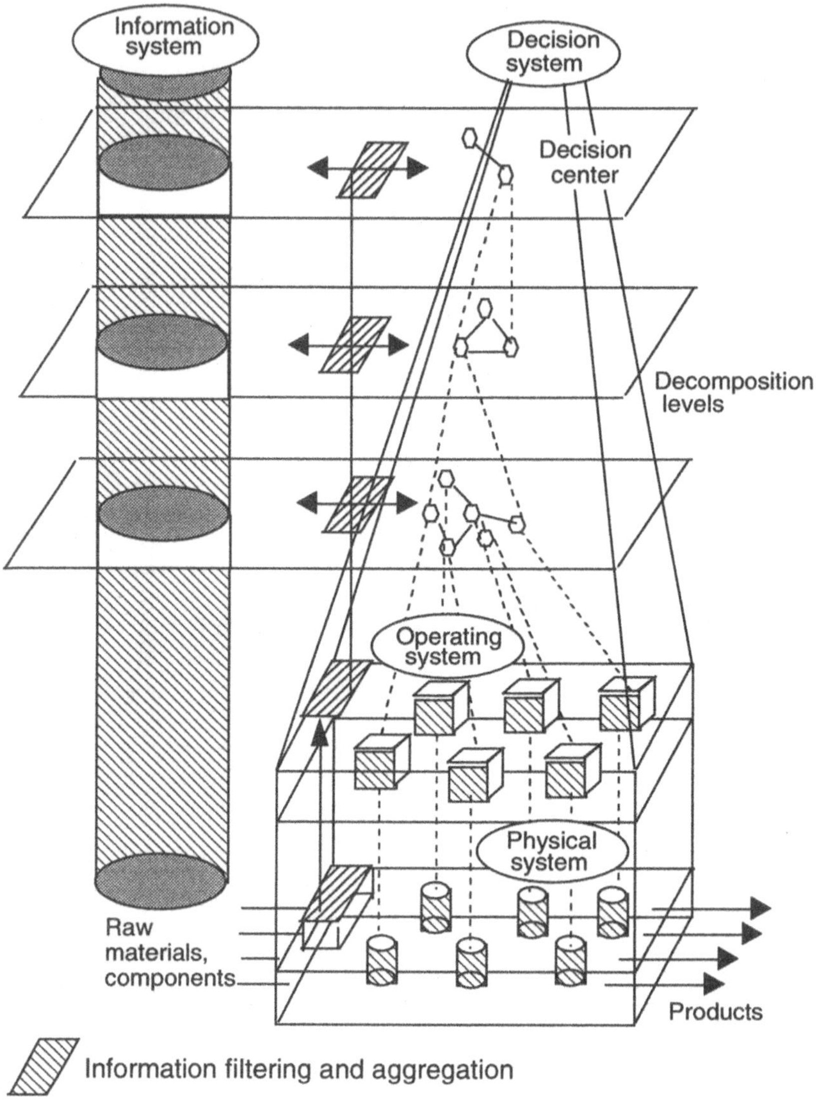

Figure 7.6 The GRAI-GIM global model.

The set consisting of the 'decision system+information system' (the control system) controls the physical system in order to reach the objectives which have been defined for the overall manufacturing system. The expression of this global action of the control system on the physical system must be carried out by each hierarchical level of the structure through the relevant decision making tasks related to the level of decision center considered. It is obvious that the decisions made by a workshop manager within his decision center must operate only on the relevant part of the physical system (here the workshop concerned). The workshop manager has a representation of the physical system completely different from the representation which the production manager has. The expression of this difference is in terms of space (a smaller domain of intervention) and in terms of time. Then for each decision center there is a specific representation of the physical system which is a more or less aggregated view of some activities contained within it.

The operational system shown by Fig. 7.6 is that part of the decision system which is closest to the physical system. The operating system is characterized by its essentially real-time nature. Considering the control system to be hierarchical leads one to consider having a corresponding hierarchical model of the physical system as in Fig. 7.7. Thus, a decision center aims at controlling a part of the physical system. The model which this part of the decision center uses to develop its control operations will come from the

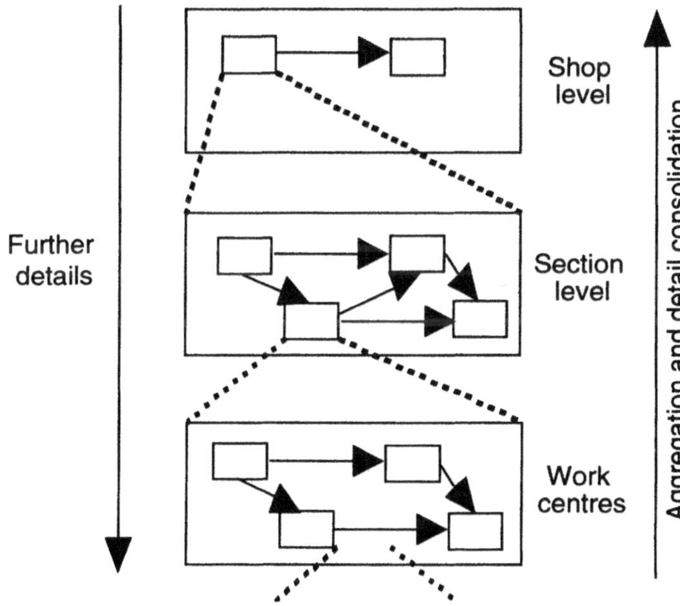

Figure 7.7 Hierarchical model of physical system.

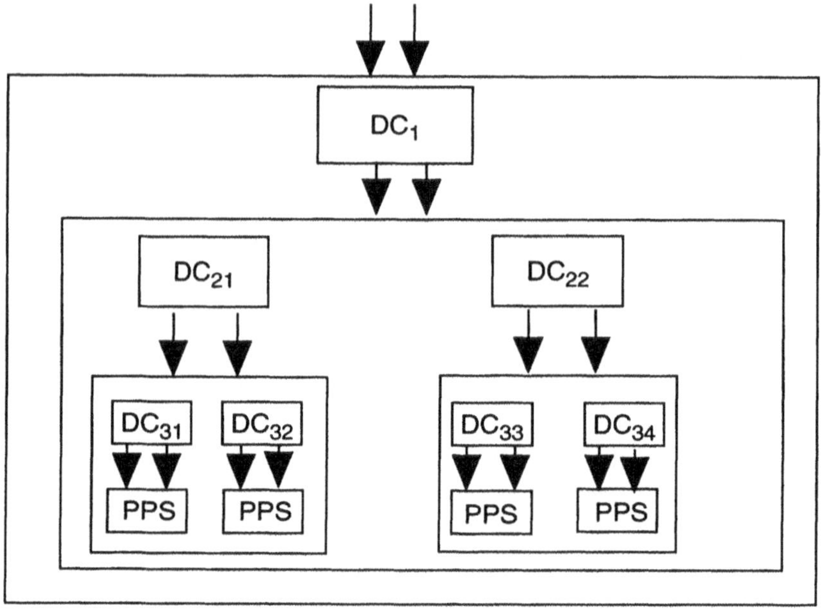

Figure 7.8 Link between hierarchical structure of the control system and hierarchical model of the physical system.

same hierarchical view of the physical system which we have just introduced. This notion leads to a structure where each decision center is dedicated to a part of the physical system as is shown in Fig. 7.8.

To complete this notion of a hierarchical system, we must go back to the concept of aggregation of information. As the level of control which is considered is raised then the information used by this particular level is aggregated or condensed in relation to lower levels. This aggregation makes it possible for this higher level to use information about a larger domain without violating any limit on the quantity of information beyond that with which it is possible to make efficient decisions on all the information available (Fig. 7.9). That is, there is a limit on the volume of information which can be considered at any level.

The domain of decision making is expressed in terms of space (the limit on the size of the part of the physical system controlled by any one decision center – Fig. 7.8) and in terms of time (the notions of horizon and period).

The detailed structure of a decision center will be described overleaf.

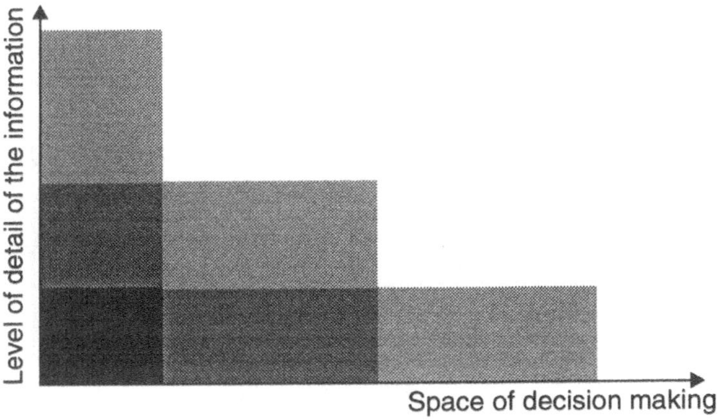

Figure 7.9 Constant volume of absorbable information.

7.2 THE DECISION CENTER MODEL

The reference model of a decision center describes the structure of the decision center itself and the links between it and its neighbors. This model also aims to describe those activities occurring within a decision center.

In order to establish the objectives of the decision center, each activity has its own purpose. These activities are organized in accordance with this purpose. This structural organization is also conditioned by the view of the physical system which the decision center has. The inner structure for these activities, essentially those which are decision oriented, is shown in Fig.7.10. The importance of this model of the physical system and the kind of information necessary for the operation of the decision center must be noticed.

In the global model, the information, decision and physical systems were split up. This division is also relevant within a decision center. Then just as the information and decision systems are connected together within the control system, a decision center is also composed of these same two systems along with a model of its physical system, the true physical system, being external to the others.

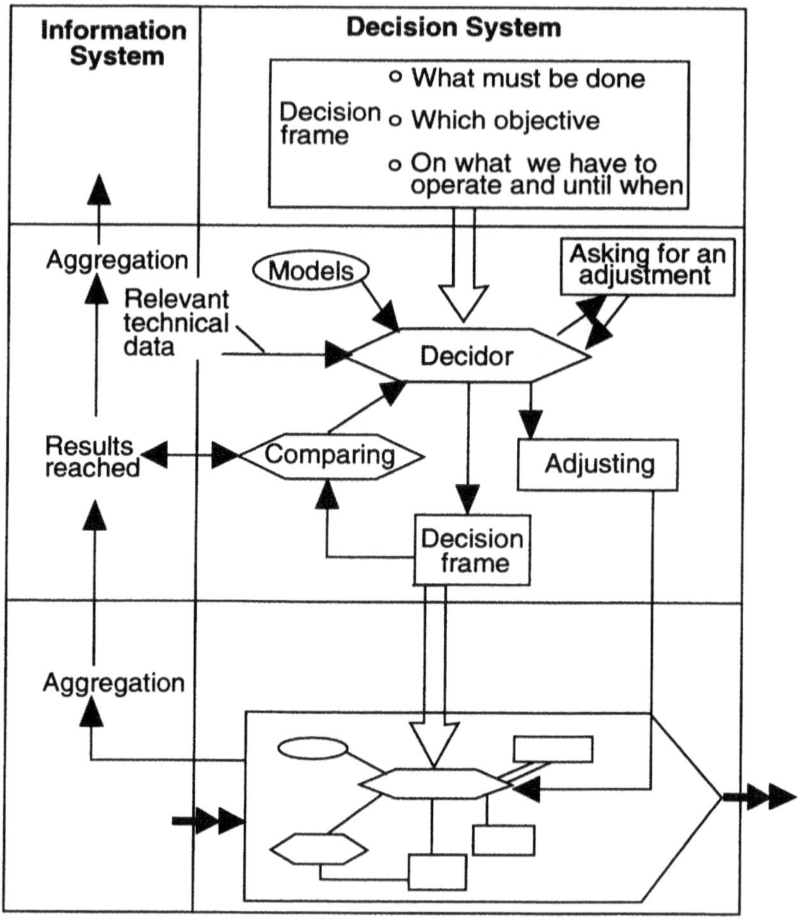

Figure 7.10 The decision centre reference model.

7.3 MODELLING FORMALISMS IN THE GRAI-GIM MODELLING FRAMEWORK

Because of the utilization by GRAI-GIM of several modelling formalisms which exist independently, this section will only present the modelling framework in which these formalisms are used and by which the consistency of the models is assured.

7.4 THE BASIC PRINCIPLE OF THE GRAI-GIM STRUCTURED APPROACH

The aim of the GRAI-GIM structured approach is to support the whole life-cycle of the manufacturing system which is to be designed. Thus, two parts of this life-cycle must be assigned a prominent position:

1. The first part which is user oriented
2. The second part which is technical.

The first part of this approach enables the users to build models establishing their own requirements for the system. The models so obtained must be readily understandable by these same users. These models are expressed according to four views: decisional, informational, physical and functional. The first three of these are related to the decomposition of a manufacturing system as introduced by the GRAI Reference Model as just described. The functional view enables the user to get a simple model of the system expressed in terms of the functions to be carried out.

The second part leads to a technical model of the manufacturing system expressed in terms of organization, information technology and manufacturing technology. Fig. 7.11 shows these two main inputs of the structured approach.

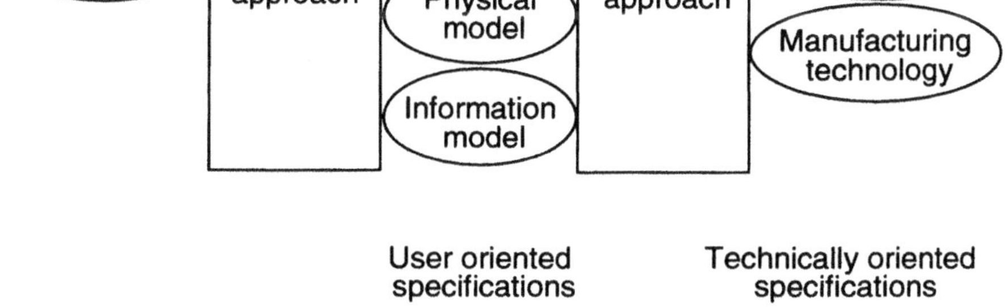

Figure 7.11 User oriented approach and technically oriented approach: the two parts of the global GIM structured approach.

7.5 THE MODELLING FRAMEWORK

Because its high complexity, an initial structuring is needed when designing a CIM system. Various concepts and models need to be defined and incorporated. In order to ensure completeness, consistency and integration between the concepts and the models, we propose to define a modelling framework in which all the models needed for the analysis, design and implementation of CIM systems can be assigned their places (see Fig. 7.12).

The modelling framework has two dimensions: functional decomposition and abstraction levels.

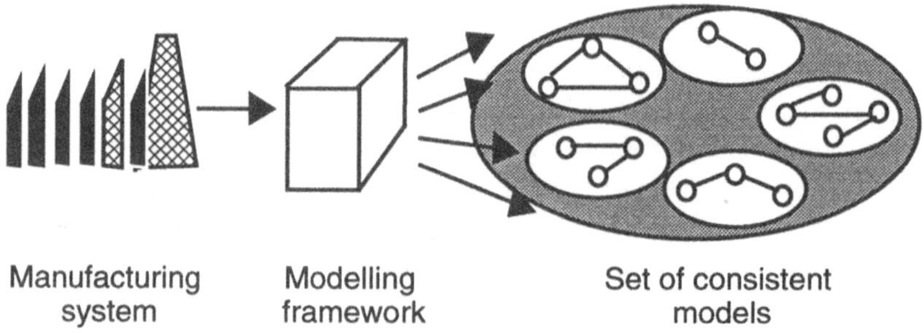

Figure 7.12 The principle of the modelling framework.

7.5.1 Functional decomposition / breakdown

As we have seen any manufacturing system may be split into three systems: the physical system, the decisional system and the informational system (Fig. 7.2). These three systems allow us to obtain three different views. A view can be defined as a selective perception of a manufacturing system which concentrates on some particular aspect and disregards others. To these three views, we add a fourth known as the functional view. The functional view enables one to get a very simple model which shows the main functions of the manufacturing system and the flows (of any nature) moving between these functions. Another advantage of these views is their ability to define exactly the boundary of the study domain.

7.5.2 Abstraction levels

The modelling activity implies a simplification of an otherwise too complex reality. Thus, a model keeps only those concepts and elements which will be necessary at the time of the model's use. The introduction of abstraction levels allows a 'stratified description' in the sense that our model is in fact composed of several, each of which integrates certain specific concepts. Practically, our model uses three abstraction levels:

1. Conceptual level: This is made up without any organizational or technical consideration, it is the steadiest level and aims to ask the question, 'What?'
2. Structural level: It integrates the organizational point of view and aims to ask the questions, 'Who?', 'When?' and 'Where?'
3. Implementable level: It is the more specific level because it integrates the technical constraints of the studied case and aims to ask the question, 'How?'

Finally the GRAI-GIM modelling framework uses the 'modelling domains' shown in Fig. 7.13.

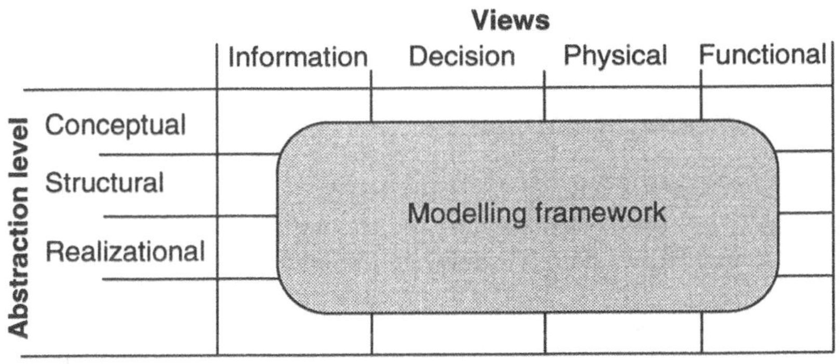

Figure 7.13 The GIM modelling framework.

7.5.3 Modelling framework and structured approach

We now need to adapt the first decomposition which was presented to include the global approach shown in Fig. 7.11 because there is a strong link between the abstraction levels and part of the approach considered (user oriented or technically oriented). In fact, the user oriented approach needs the upper abstraction levels and the technically oriented approach needs the

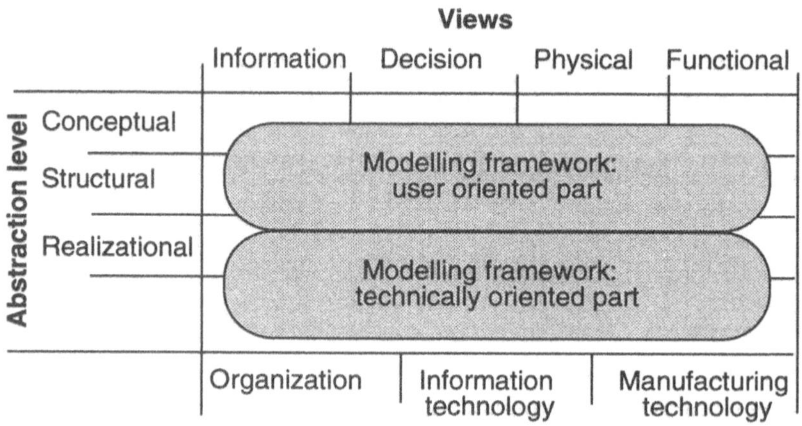

Figure 7.14 The GIM modelling framework: taking into account. the methodology.

lower ones. Fig. 7.14 shows more precisely the modelling framework which must be taken into account in using these two approaches.

7.6 MODELLING FORMALISMS

This discussion will focus on the upper levels of the modelling framework (those which are user oriented). Thus, we will now see the eight corresponding modelling domains (formed by the intersection of a column and a row of the diagram of Fig. 7.14). The formalisms used for each subdomain must be well adapted to the modelling needs of each of them. The needs for each subdomain are as follows:

7.6.1 Conceptual information model

The Conceptual Information Model is a description of all of the stable and 'natural' data of the organization, of their attributes and of the links between them. The formalism used here is the Entity Relationship model.

7.6.2 Structural information model

The Structural Information Model describes the data structure in relation to the distribution of data and the computerized/manual choice. The formalism used here is also the Entity/relationship model.

7.6.3 Conceptual decision model

The Conceptual Decision model is a description of the decision taking structure, the links between decision levels, an analysis of the links between objectives, an analysis of the constraints, and a description of the decision variables involved. The formalism used here is the GRAI grid at the global level and the GRAI net at the detailed level.

7.6.4 Structural decision model

The Structural Decision Model mainly enables the identification of decision makers, and their responsibility and authority. It links decision makers and the decision making process. The formalism used here is the GRAI grid at the global level and the GRAI net at the detailed level.

7.6.5 The Conceptual physical model

The Conceptual Physical Model is a description of the process and the routes taken by the physical flows between operations. The formalism is the actigram IDEF0.

7.6.6 Structural physical model

The Structural Physical Model gives information about time, work centres and operators. It includes the elements about linking and synchronization and indicates who does what. The formalism is the actigram IDEF0.

Fig. 7.15 shows the user oriented part of the GRAI-GIM reference framework with the modelling formalisms used.

7.7 VALIDATION BETWEEN MODELS

Validation between models is one of the foundations of our approach. We began indeed by applying an analytic approach (decomposition) to the studied system. We went from an overall view of the system/real object to these

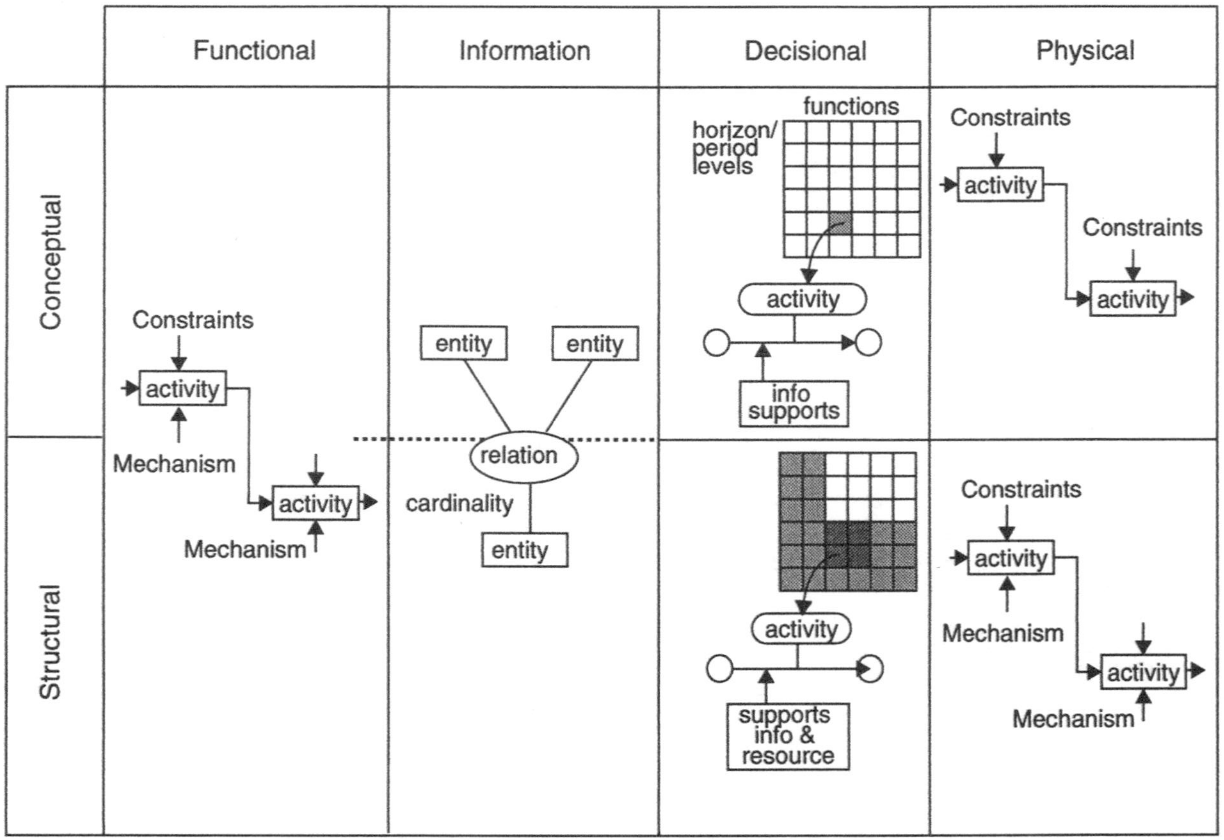

Figure 7.15 GIM reference framework with the modelling formalisms used.

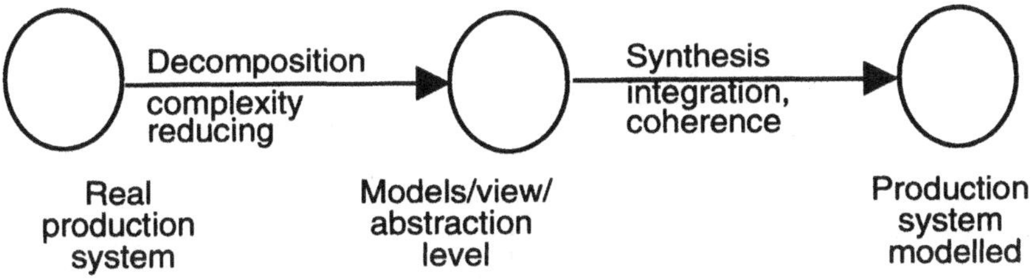

Figure 7.16 Decomposition / synthesis type cycle supporting the GIM approach.

separated models according to the views and abstraction levels described. The complete model will be obtained only by applying a set of validation rules between each set of the model. The aim of the validation is to impose a global coherence onto the study. The structured approach is supported by the application of a decomposition/synthesis type cycle (Fig. 7.16).

Validation between models can be achieved only if we can find an element present in all the partial models. As shown on Fig. 7.17 the subsystems are connected through data:

1. The decision subsystem makes decisions and processes data
2. The physical subsystem aims to process physical flows but is the source of process data for the information and decision subsystems.

Data are the only means of communication between subsystems, and they are the validating elements required. Thus, the basic point of the validation between models will be realized by imposing data coherence. Fig. 7.18 shows an example of validation between the Conceptual Decision Model and the Physical Conceptual Model through data.

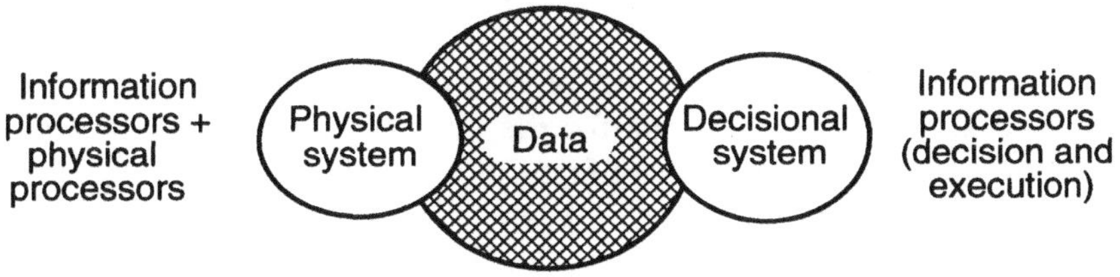

Figure 7.17 Data as vector of communication.

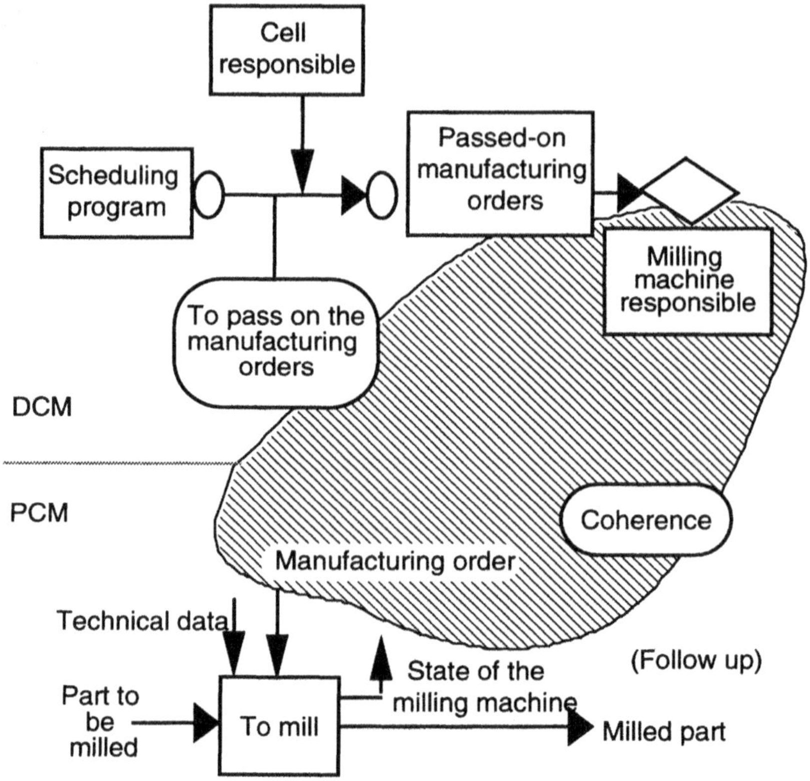

Figure 7.18 Example of validation between conceptual decision model and physical conceptual model through data.

7.8 GRAI-GIM STRUCTURED APPROACH

According to the user requirements for the future system, the goal of GRAI-GIM is to provide specifications for the system in terms of organization, information technology and manufacturing, which will allow the user to build this new system.

The structured approach of this method has four major phases: initialization, analysis, design and implementation. We will focus mainly on the three first phases and will provide only some 'orientations' for the implementation phase. See Fig. 7.19.

One of the main advantages of this method is to split up the design phase into two subphases: that for the user oriented design and that for the

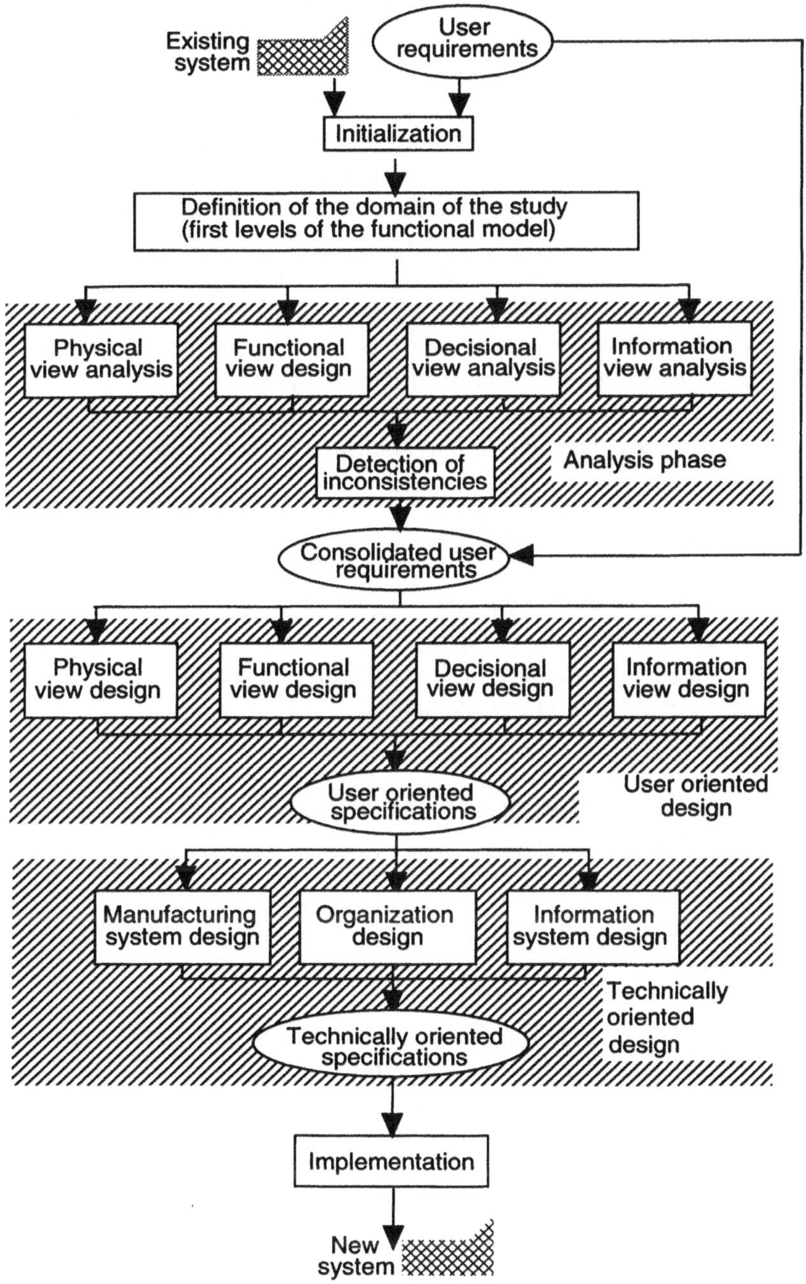

Figure 7.19 GRAI-GM structured approach.

technically oriented design. We consider that the techniques used to build new manufacturing systems are currently very difficult for the users of the future system to understand. This is particularly true for the information technology domain. In addition, because of the amount of investment necessary to build a new manufacturing system, we need to be sure that the design of the new system reaches the objectives defined in the user's requirement. The design of the new system must thus be validated by the user before the start of any developments or implementations.

Because of these remarks, we have split up the design phase into two subphases: the user oriented design which provides a set of user oriented specifications understandable by the users and the technically oriented design which provides the technical specifications necessary for the development and the implementation of the future system.

The users will then validate the user oriented specifications to be sure that the design provides the efficient solutions to their requirements.

7.9 INITIALIZATION PHASE

This phase includes eight main steps:

1. Overall definition of the company
2. Presentation of the method to the company managers
3. Definition of the domain of the study
4. Diagnosis
5. Initialization milestone
6. Definition of the people involved in the study
7. Training of the participants
8. Planning and resources for the next phase.

7.10 ANALYSIS PHASE

This step permits us to define the characteristics required of the current system. It is also necessary to know the interfaces between the studied domain and other functions like Marketing or the Financial Department for example.

As it has been defined in the GRAI-GIM modelling framework, the analysis concerns mainly the modelling of the four views: functional, decisional, informational and physical views. To model these views, we will use the modelling tools defined previously for the design of the GRAI-GIM architecture (IDEF0, GRAI grid and nets, and the Entity Relationship tools).

7.11 DESIGN PHASE

When the analysis phase has been approved as noted above, it is necessary, before starting the technical design, to redefine clearly the various inputs which will be the basis for this phase. We have formalized the existing system, we have detected the inconsistencies of this system and we know the objectives of the company. The design phase is split up into two main steps. A first step as noted is more user oriented where the users are involved during the design of the four views and must validate the results to allow the study to continue. The result of this first step is the user oriented specifications. Working according to these specifications, experts in manufacturing, organization and information technology are then put in charge of performing the technically oriented design step. The result of this step will be the technical specifications of the future system which will allow the user to choose system components (machines, computers, software packages, databases, human resources, etc) from the market or to develop them themselves, and then implement the new system.

7.12 IMPLEMENTATION PHASE

The implementation phase can itself be split up into three tasks: the installation of the new resources and the training of the users, the preparation for the starting up of the new system and finally, the complete operation of the new system. The problem in this phase is to ensure a good transition between the current and the new system with a minimum of impact on the production operating.

7.13 A SUMMARY OF HOW TO USE GRAI-GIM

The objectives of GRAI-GIM are to develop the specifications for a CIM system in order to:

1. Design the specific architecture of the CIM system being studied with all the components and the links between them, and
2. To give the specifications for all these components in order to select them from those already on the market, or to define the requirement for the development of these components.

If we consider the life-cycle of the manufacturing system, GRAI-GIM concerns itself only with the first phase: Analysis/Design, it does not treat

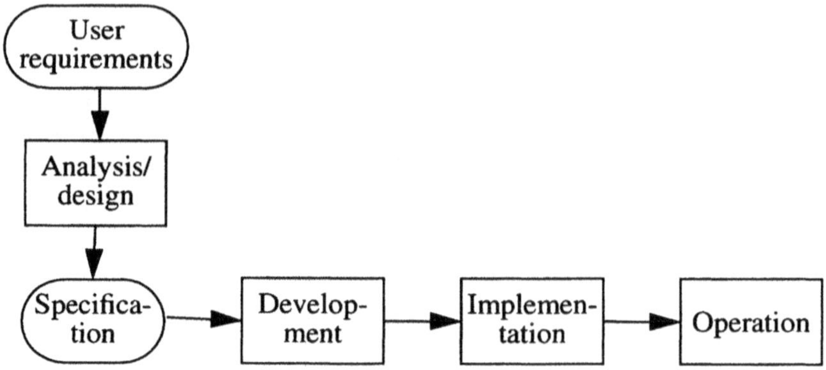

Figure 7.20 Life-cycle of a manufacturing system.

Development. GRAI-GIM is an input for this second phase as shown in Fig.7.20.

GRAI-GIM could be helpful in the Implementation Phase or in the Operating Phase as a reference, as a guide, or a training tool.

The difficulties involved in producing these specifications are the consequences of a double set of constraints:

1. The specifications for the proposed solutions must be validated by the users
2. The specifications for the proposed solutions must be understood on a technical point of view by the potential vendors of the necessary components.

To answer to this dual problem, GRAI-GIM proposes to develop two kinds of specifications:

1. User oriented
2. Technically oriented (see Fig. 7.11).

For the user oriented specifications, the GRAI Model is used with the three subsystems: decisional, informational and physical plus the functional view. We describe these specifications at each of the two levels:

1. Conceptual: in term of concepts,
2. Structural: in term of structure.

Taking the model for the Physical subsystem for example, at the conceptual level, one defines the basic manufacturing operations and the global capacity needed for each type of manufacturing operation (turning, drilling, milling, etc.). The physical model at the structural level evaluates the number of machines needed, and the placement of these machines into cells or shops. Various criteria must be considered for this placement, such as economic criteria, safety criteria and flexibility criteria. For example, to structure three manufacturing operations (turning, drilling and milling) into a cell, one can use three individual machines or one machining center capable of performing all the three kinds of operations. The first structure is safe and allows one to reach to a machine breakdown, the second structure is flexible but is less safe.

For technically oriented specifications we have analyzed the potential groupings of solutions and have classified these solutions into three domains:

1. Information Technology: hardware, software, packages, (CAD systems, Computer Aided Production Management, Computer Aided Maintenance, Computer Aided Process Planning....) Local Area Networks, Data Bases Management Systems, Controls for NC machine tools, Robots, transport systems and warehouses.
2. Manufacturing Technology: all equipments concerning Machine Tools, Specialized Machines, Warehouses, Transport systems, Robots, etc., as well as Human Operators.
3. Organization: Concerned mainly with how the management of the manufacturing system is organized, the principles of management, the organization into workshop, cell....

The technically oriented specifications are involved with two levels: structural and implementable. At the structural level we give the generic requirements in order to choose the solution. At the implementable level, we choose an effective solution available on the market. At this last stage, we may find that the solution does not exist on the market and a new device must be developed. The specifications are the starting point for this development.

7.14 CONCLUSION

The GRAI-GIM method is developed more for the user oriented design than for the technically oriented design. The design of the GRAI-GIM method is an on - going process and the next development will allow an improvement of the technically oriented design phase. Besides, because of the amount of work necessary to model all of the necessary views used during the analysis and the design phase, the use of a computer tool support is absolutely necessary. All the supports allowing the design to be concentrated on added value activities must be provided. A computer tool support will allow one to facilitate the modelling of the views, the coherence checking between the models, the detection of inconsistencies, the use of reference models during the design phase and the production of documents.

7.15 REFERENCES

[1] Doumeingts, G., Methode GRAI: Methode de Conception des Systemes en Productique, Thesis d'etat: Automatique: Universite de Bordeaux I, pp. 519 (November 13, 1984).
[2] Doumeingts, G., Vallespir, B., Darracar, D. and Roboam, M., 'Design Methodology for Advanced Manufacturing Systems,' Computers in Industry, Vol. 9, No. 4, pp 271–296 (December 1987).
[3] Doumeingts, G., Vallespir, B., Zanettin, M., and Chen, D., 'GIM – GRAI Integrated Methodology, A Methodology for Designing CIM Systems,' Version 1.0, Unnumbered Report, LAP/GRAI, University Bordeaux 1, Bordeaux, France (May 1992).

8

An overview of PERA and the Purdue Methodology

T.J. Williams

In this short chapter we will present only an overview or eagle's eye view of this Architecture and its meaning and the gist of its various ramifications and uses without describing all of its details. For this purpose we will concentrate our early discussion on a computer integrated manufacturing (CIM) system in a factory making relatively common metal products. This is the easiest application of the Architecture to describe. The later part of this discussion will then show the extension of this Architecture to cover any type of Enterprise endeavor and not just CIM in the factory.

The Architecture is fully described in [1].

8.1 A STRUCTURE FOR THE PURDUE ENTERPRISE REFERENCE ARCHITECTURE

Fig. 8.1 presents the skeleton diagram or framework against which all of the work carried out with the Architecture can be discussed. This structure, progressing from top to bottom, represents the life history of the enterprise integration project from its initial concept through the stages of functional analysis, functional design or specification, detailed design, construction and installation to operation and finally to obsolescence. There is only a pseudo time scale involved in this since no exact time dimension is implied. However, most, if not all, projects progress through these various stages.

As will be readily seen in the discussion to follow, the several steps and blocks indicated in the progression through the structure of Fig. 8.1, provide important information regarding the work to be done at each step or block in developing the CIM system. They can also be used to indicate at each step or block the project aids available; applicable modelling or analysis techniques which can be used; example systems in the literature which would be helpful illustrations at that point; and a great deal of other

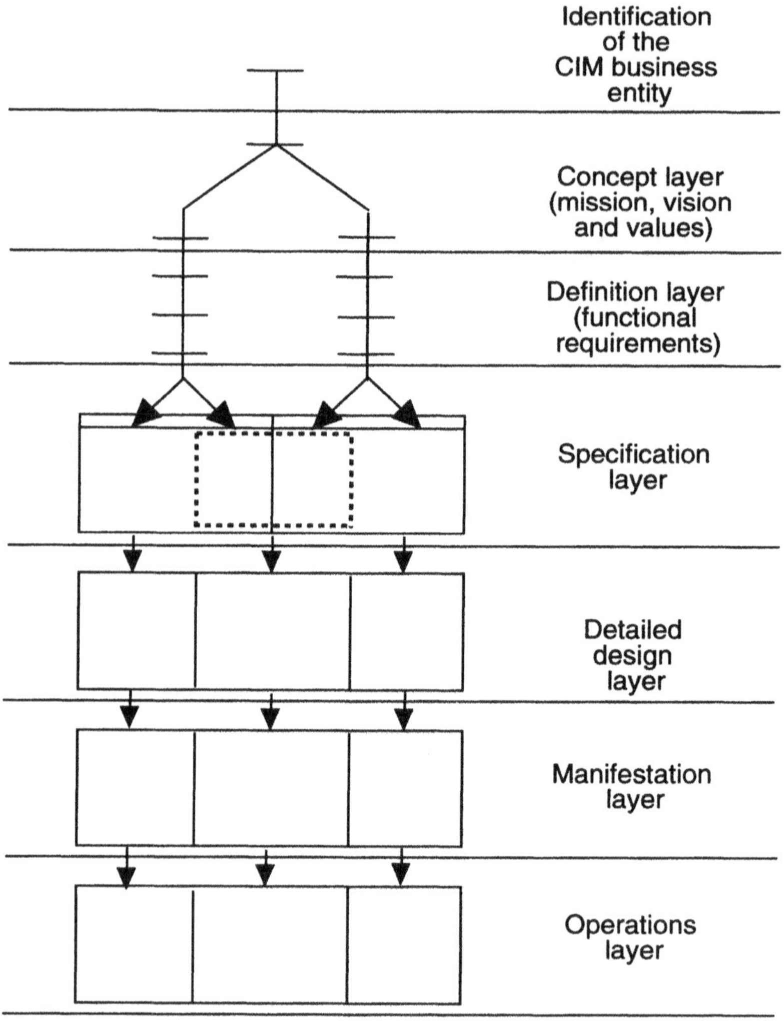

Figure 8.1 A layering of the Purdue Enterprise Reference Architecture in terms of the types of tasks which are occurring within those regions on the graphical representation of the architecture.

important information for the practitioner of enterprise integration. Much of the content of the referenced text [1] involves an explanation of what is done at each step or block in the architecture and how that relates to the work of adjacent steps or blocks.

8.2 DESCRIPTION OF THE CONCEPT AND DEFINITION LAYERS OR PHASES

Note that the structure begins at the top with the identification of the CIM Business Entity, the factory or section of plant in which the contemplated enterprise system is to be installed. We then proceed to develop the Concept Phase or Concept Layer. Any enterprise integration program must be business driven and reflect the goals and aspirations of management concerning the expected outcome of the integration endeavor. Therefore the first requirement is to establish management's mission, vision, and values for the program as expressed in their goals, objectives, mandates, etc., concerning the program. This is shown in Fig. 8.2.

Management's requirements as expressed by the mission, vision and values and related statements are then converted into policies concerning the design, development, construction and operation of the completed factory manufacturing integration unit. The resulting policies, which complete the Concept Phase, are converted into a set of Functional Requirements which define the tasks which our proposed integration program must be able to carry out when completed in order to fulfill the mission, policies, etc., already expressed by management.

These requirements are then voiced in the form of a set of elementary, modular tasks which when implemented will satisfy these requirements. These elementary tasks are then connected together to form a set of functional modules and finally the functional modules themselves are further connected into networks.

Since there are only two basic types of tasks which can be carried out in the manufacturing facility:

1. Those related to the physical manufacturing operations themselves, and
2. Those related to information concerning the manufacturing operations and their control, i.e. sensor readings, control commands, scheduling information, production data, etc.

There are two and only two streams required to define the functional requirements of the proposed system. We shall call these respectively the information (left hand) and manufacturing (right hand) side of the Architecture up to this point.

We have used networks as the method for graphically displaying the functional requirements since these are the popular mode today. On the information side these may become data flow diagrams [2,3], IDEF0 diagrams [4,5] or entity relationship diagrams [6,7]. Many other graphical and computer based schemes have also been proposed. On the manufacturing

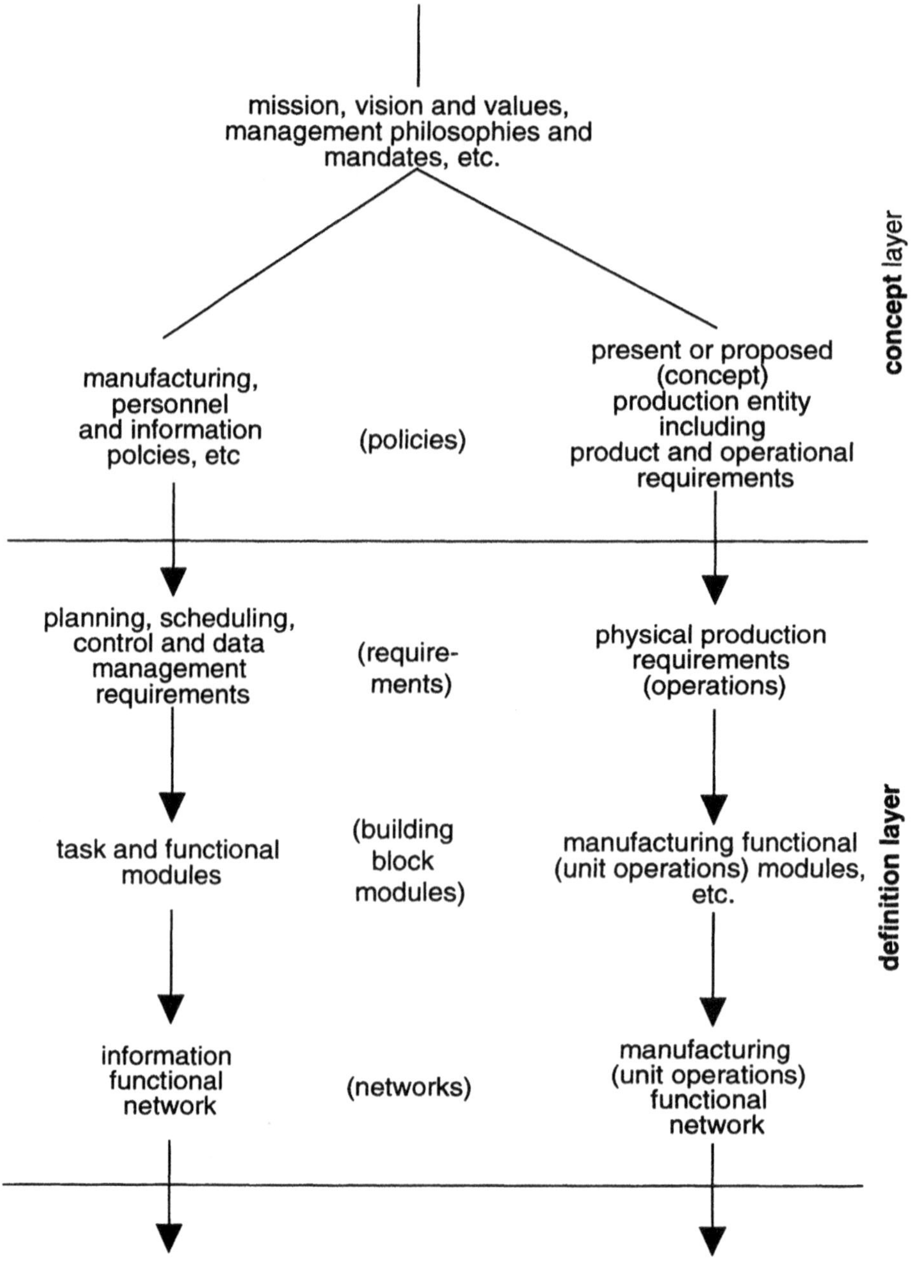

Figure 8.2 Definition of the components of the concept and definition layers.

side these networks are commonly expressed as process flow diagrams, and material and energy flow diagrams. The development of the modelling of the tasks and functional modules and their collection into macromodules and networks on both sides of the Architecture are thoroughly covered in Chapter 3 of the referenced text [1]. The resulting networks, if sufficiently detailed, will satisfy the functional requirements definition for the CIM system for both types of tasks and functions required and therefore for both sides of the Architecture.

Once the functional requirements are satisfied we are ready to consider how we will implement these requirements. Note that up to this point, this has not been a consideration. Provided all functional requirements are met, it makes no difference what method of implementation is used, i.e. it makes no difference whether they are conducted by humans or machines (or in what type of equipment or where). All of the latter considerations are implementation details. Therefore, as carried out here, any discussion of the place of the human can be postponed in the manufacturing system development until after all tasks and functions are defined.

The range of human skills, muscle power, and other measures of human capability as well as the corresponding abilities of mechanical and electronic devices are such that not all of the tasks in any manufacturing facility can be carried out by either humans or machines alone. Thus we must provide for both humans and machines in carrying out both the information requirements and the manufacturing requirements.

Remember that we have defined all tasks in a modular and functional fashion. This modular and functional character remains regardless of how they are implemented at the moment. Thus, if one should wish to convert any part of the system later in the life of the project from human to machine for implementation or vice versa, the work required should involve only the definition of the alternate implementation and the provision of the proper interfaces to connect the new implementation means into the prevailing system to accomplish the transition.

The endeavor now facing us is to convert the two task streams (information and manufacturing) into three implementation streams (comprising respectively, computer and other control system elements, humans, and manufacturing equipment). Since each of these new streams will eventually have a physical manifestation it is important at this point to decide how to name them.

8.3 A SYSTEM OF NOMENCLATURE

In the manufacturing and enterprise modelling disciplines the word architecture has two different, although related, meanings. Many architectures, including most of those referenced in the literature, are, like this one, a structure or reference framework by means of which one can discuss many different aspects of topics concerning the CIM system or enterprise involved. The other use of the word, relating more to its traditional meaning, concerns the physical structure of the system we are considering. How do the parts connect together, etc.? This latter is especially common in discussing the physical organization of computer and communications systems, for example.

The three streams we are now going to consider involve the use of both definitions of the word, architecture, as defined above. With apologies to the reader for any confusion it may cause, we would now like to define not three, but five subarchitectures of the Purdue Enterprise Reference Architecture. The first two of these are shown in Fig. 8.3. They represent merely the one-for-one implementation of the functional tasks already discussed without yet defining the means of implementation. These will be important to our later discussions, hence their definition here.

However, as we have already noted, some of the tasks in both functional streams must be implemented via human activities. Therefore the structure of Fig. 8.3 must be modified as shown in Fig. 8.4. Part of the tasks of the Information Architecture are carried out by computers and other electronic, pneumatic, or mechanical devices (these form the Information Systems Architecture). Others are carried out by humans (these form part of the Human and Organizational Architecture). Note again that it is the functional tasks that are being distributed here. Therefore the functional definition has not been changed. Only their method of implementation has been decided.

Likewise on the manufacturing side, part of the tasks of the Manufacturing Architecture are carried out by physical equipment (these form the Manufacturing Equipment Architecture). Others are carried out by humans (these form the remainder of the Human and Organizational Architecture). Again, as before, these are functional tasks that have been distributed while their functional definitions have not changed. It is very important to make this distribution at this stage since the actual implementations of the resulting three important implementation architectures are so vitally different from each other from here on out in our discussion.

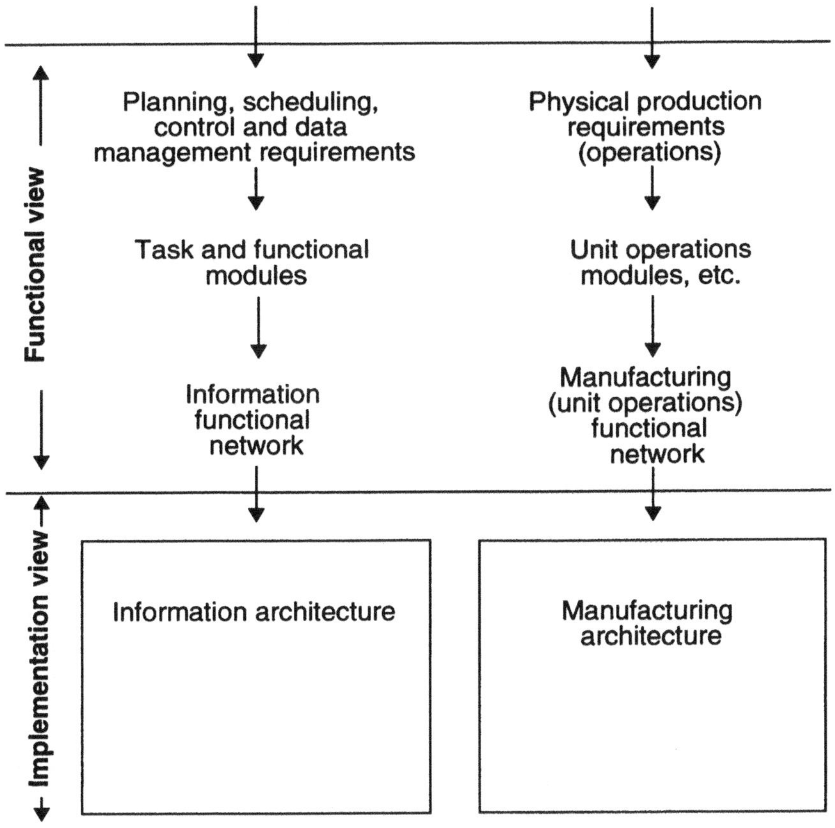

Figure 8.3 Developing the relationships of the several subarchitectures of the Purdue Enterprise Reference Architecture for manufacturing systems.

If no humans were involved, the Information System Architecture would fill the whole width of the Information Architecture block. Likewise the Manufacturing Equipment Architecture would fill the whole width of the Manufacturing Architecture. The result would be the so called 'lights out' plant, i.e. completely automated.

Therefore the dashed lines in Fig. 8.4 separating the Human and Organizational Architecture from the other two implementation architectures represent the degree of automation used in making the decision as to where specific tasks fall between the three architectures. These lines can be defined as shown overleaf.

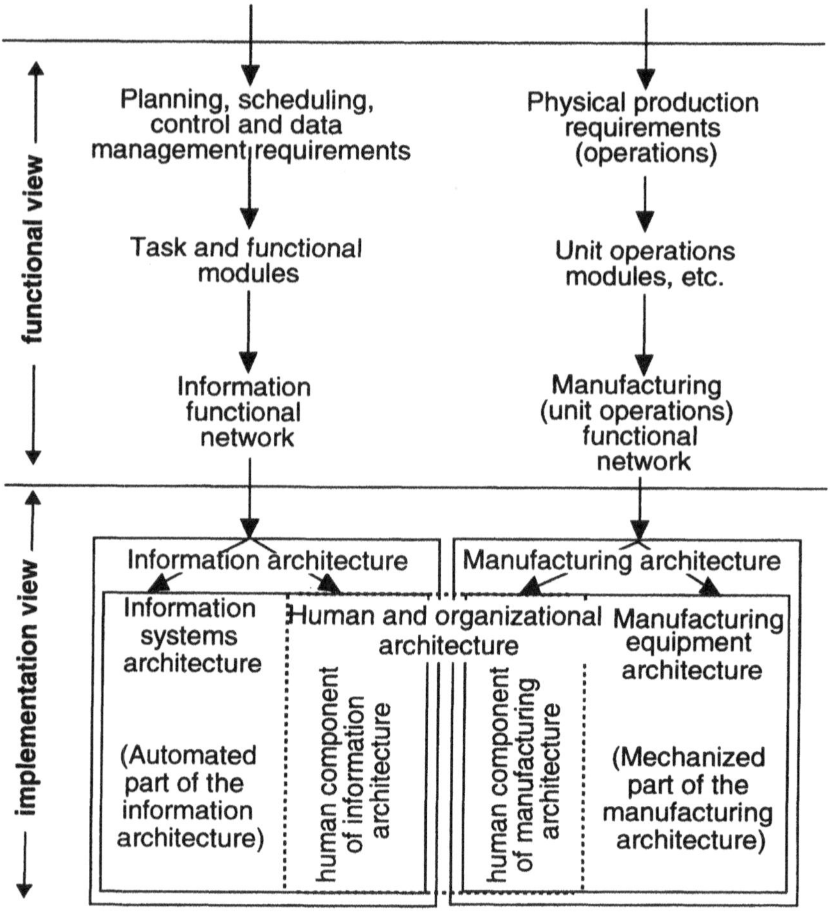

Figure 8.4 Division of the implementation view into the three implementation architectures.

8.4 DEFINITION OF THE PLACE OF THE HUMAN AND ORGANIZATIONAL ARCHITECTURE

There is a line which can be called the Automatability line which shows the absolute extent of technology in its capability of actually automating the tasks and functions of the manufacturing system of the CIM Business Entity. It is limited by the fact that many tasks and functions require human innovation, and cannot be automated with presently available technology (see Fig.8.5).

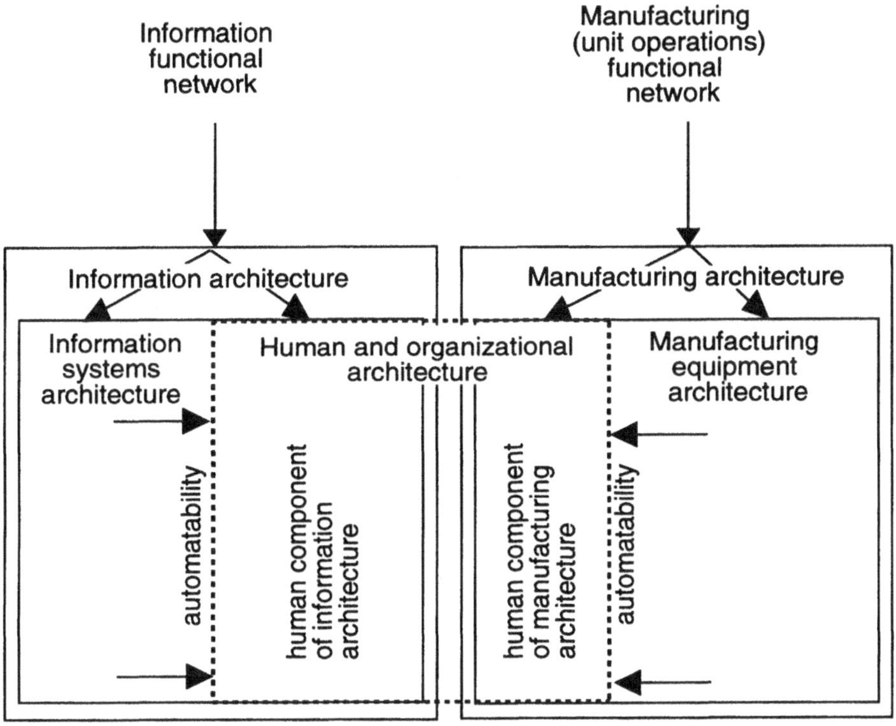

Figure 8.5 Definition of the human and organizational architecture showing the automatability line.

There is another line, which can be called the Humanizability line, which shows the extent to which humans can be used to actually implement the tasks and functions of the integration system of the CIM Business Entity. It is limited by human abilities in speed of response, breadth of comprehension, range of vision, physical strength, etc. Of course, prior to the industrial revolution most information functions in manufacturing were human implemented (see Fig. 8.6).

There is still a third line which can be called the Extent of Automation line (Fig. 8.7) which actually defines the boundary between the Human and Organizational Architecture and the Information Systems Architecture on the one hand, and between the Human and Organizational Architecture and the Manufacturing Equipment Architecture on the other. The Extent of Automation line shows the Actual Degree of automation carried out or planned in the integration system of the CIM Business Entity.

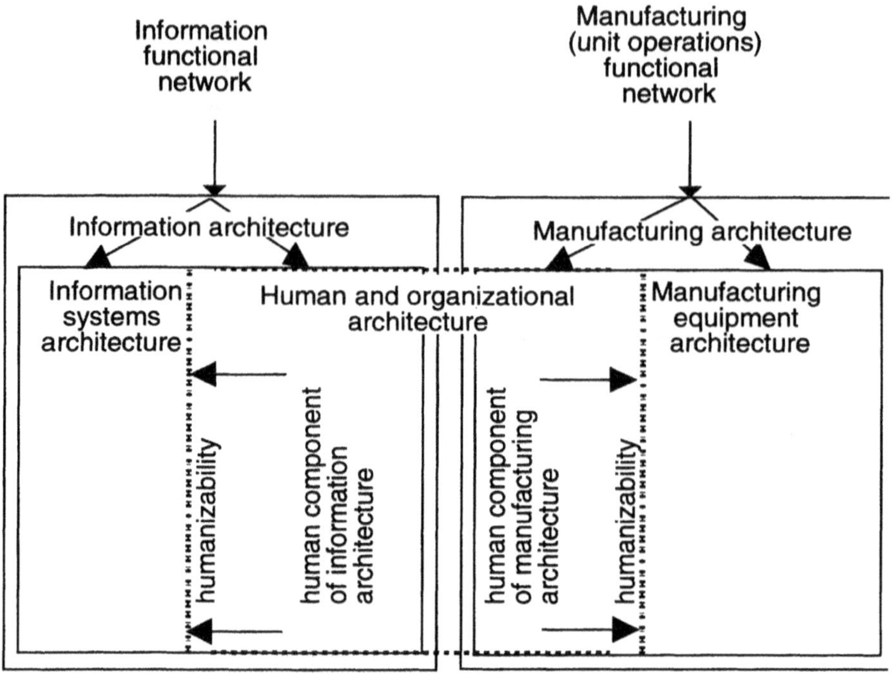

Figure 8.6 Definition of the human and organizational architecture showing the humanizability line.

The location of the Extent of Automation line has:
1. Economic
2. Social (Customs, Laws and Directives, Union Rules), as well as
3. Technological

factors in its determination. This is the line actually implemented.

An Automatability line showing the limits of Technology in achieving automation will always be outside of the Extent of Automation line with respect to the automation actually installed. That is, not all of the technological capability for automation is ever utilized in any installation for various reasons. Thus the Human and Organizational Architecture is larger (i.e. more tasks or functions) and the Information System and Manufacturing Equipment Architectures are smaller (fewer functions) than technological capability alone would allow or require.

The Human and Organizational Architecture is implemented mainly through the answering of human relations concerns; such as the level of skills required, the training involved in establishing and maintaining the

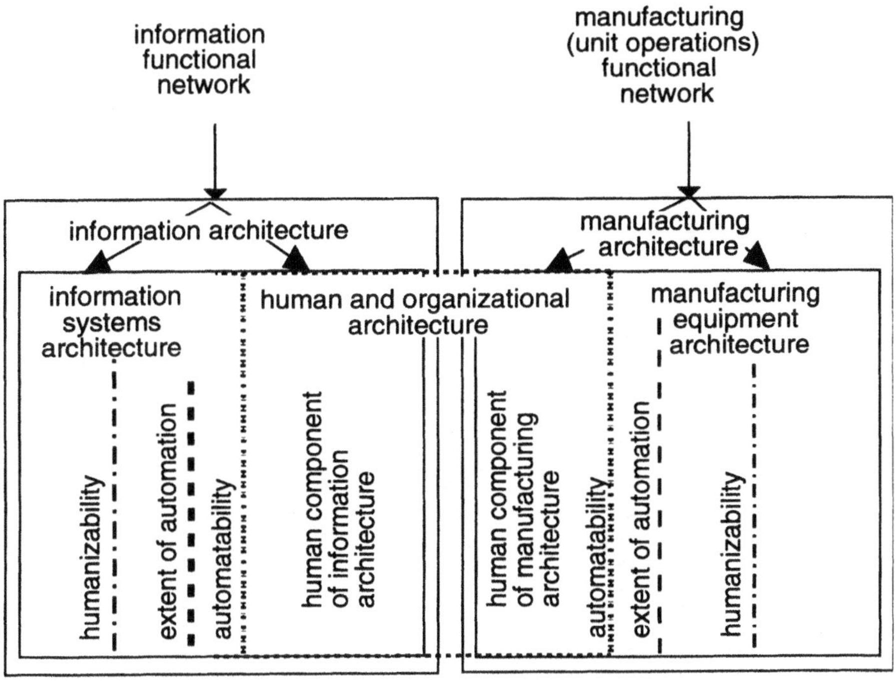

Figure 8.7 Reations of the automatability, humanizability and extent of automation lines in defining the human and organizational architecture.

appropriate skill levels; union concerns such as assignment of personnel to tasks, crafts involved, etc.; such things as the organization used and the resulting reporting paths, pay and vacation, and so forth.

The Information Systems Architecture will involve computer and communications systems choices, database techniques, programming languages, as well as the choice of application programs to satisfy the tasks. The Manufacturing Equipment Architecture will involve the design or procurement of the necessary machines and other manufacturing devices, their appropriate layout in the plant, and the provision for movement of raw materials and semifinished and finished parts about the plant.

Thus again, if the functional requirements are satisfied and if the appropriate coordination is maintained, the implementation of these three architectures can proceed relatively independently of each other. This is a major benefit of this viewpoint particularly when their individual needs are so different.

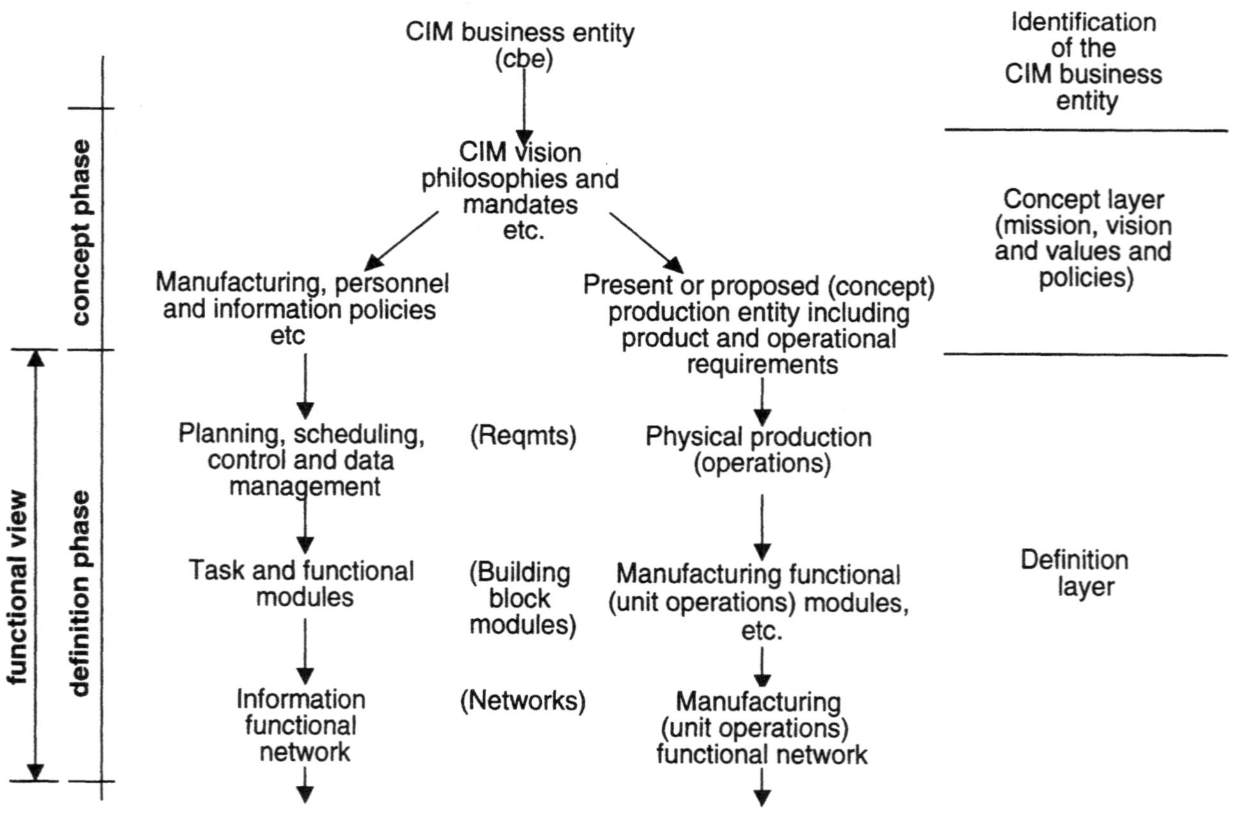

Figure 8.8 (a) Development of a CIM program as shown by the Purdue Enterprise Reference Architecture.

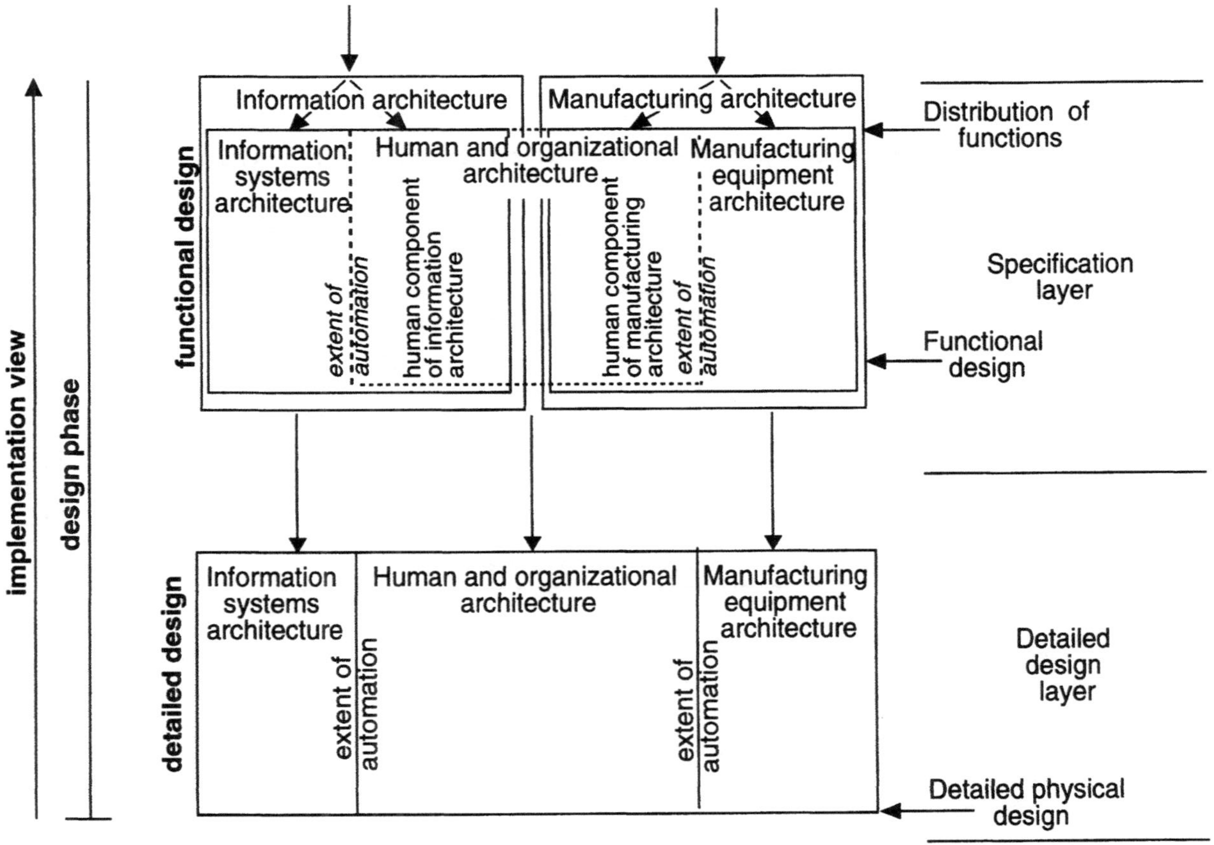

Figure 8.8 (a) Development of a CIM program as shown by the Purdue Enterprise Reference Architecture (Continued).

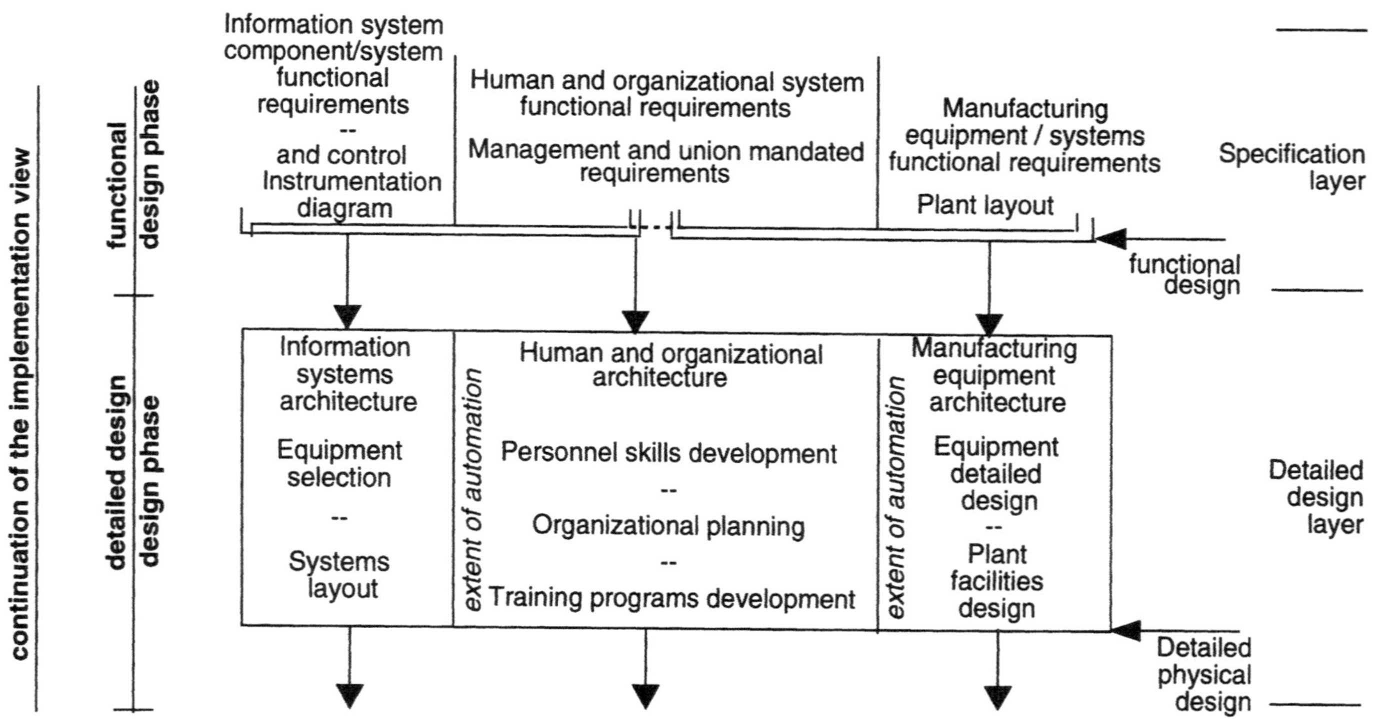

Figure 8.8 (b) The later phases in CIM system evolution and their tasks in relation to the Purdue Enterprise Reference Architecture

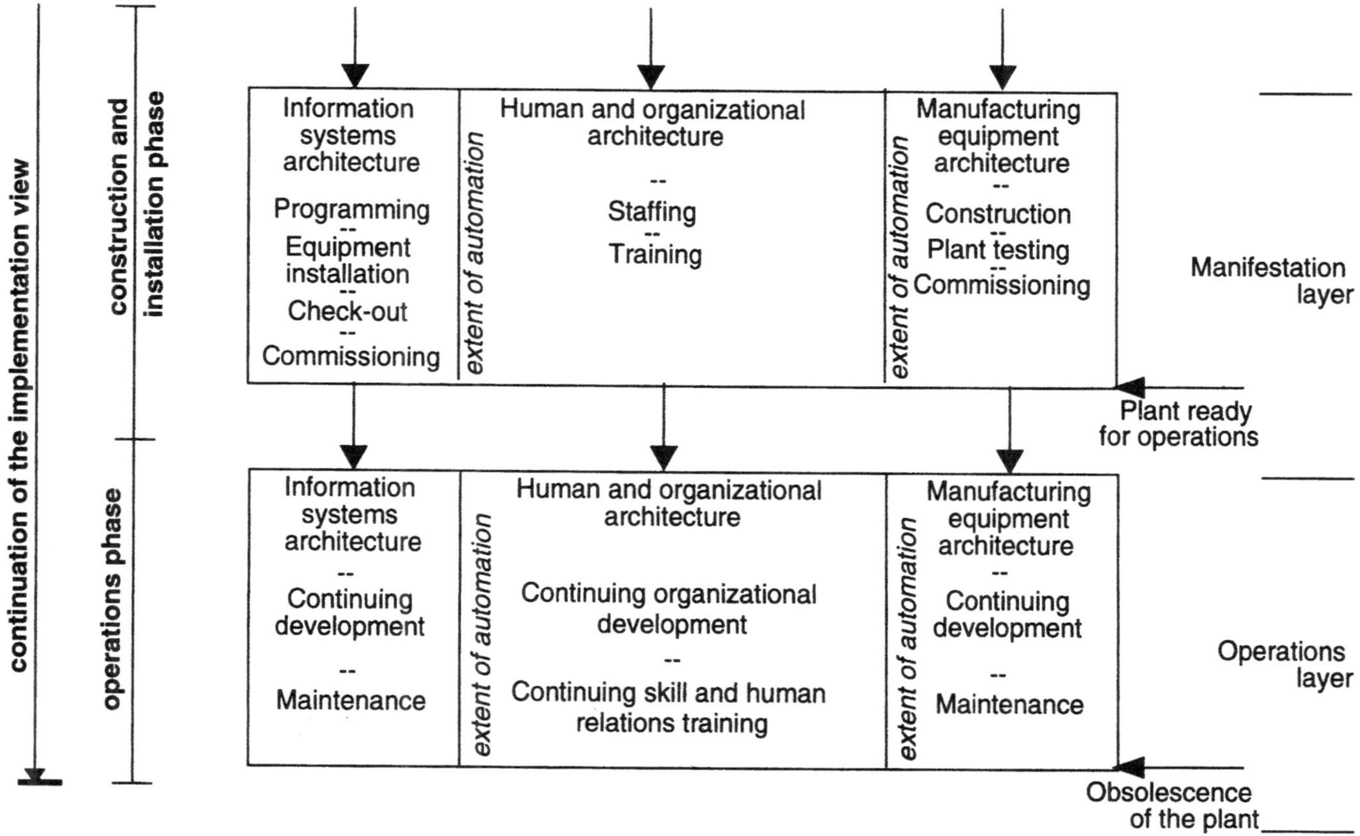

Figure 8.8 (b) The later phases in CIM system evolution and their tasks in relation to the Purdue Enterprise Reference Architecture (Continued).

8.5 DEVELOPMENT OF THE IMPLEMENTATION OF THE CIM SYSTEM

As shown in Fig. 8.1, implementation of the manufacturing integration program as outlined by the Architecture proceeds from Functional Design or Specification through Detailed Design to Construction, Installation and Check-out and finally to full Operation which continues to Plant Obsolescence.

Figures 8.8a and 8.8b show the fully annotated structure of the architecture, indicating at each layer the types of operations which are carried out concerning each task involved.

8.6 SUMMARY OF THE OVERVIEW AS RELATED TO MANUFACTURING

As mentioned at the start of this discussion, the straightforward manufacturing implementation as discussed here is the easiest class of applications of the Architecture to describe and was therefore used here to acquaint the reader in general with the description of the use of the Architecture. The explanation of much more complex examples will be brought up next in this discussion.

8.7 ENTERPRISES AND THE ARCHITECTURE

Enterprises are organizational entities built to produce goods and/or services in response to customer needs [8]. Enterprises are highly dependent on market trends and needs and customer satisfaction. Enterprises must face two dynamic environments:

1. The external environment characterized by rapidly changing market needs and competition from other vendors serving the same field of endeavor.
2. The internal environment characterized by global objectives, management attitudes, and technological changes.

The Purdue Enterprise Reference Architecture therefore views the enterprise from two distinct viewpoints, that of the business environment as seen by the business user and that of the enterprise's resources (technology of the field of endeavor and information technology). These two descriptions are reflected in the functional requirements modelling layer (Definition Layer), in the functional specification modelling layer (Specification Layer) and in the detailed design modelling layer (Detailed Design Layer). The latter two are implementation layers. See Figures 8.1 and 8.8a and 8.8b.

Enterprises may also be defined as entities which exist to carry out a mission or missions. The mission involves one or more tasks carried out with the aid of resources to satisfy one or more desired ends (goals or objectives) in the face of difficulties and limitations (constraints). Note that all of the above applies to every enterprise regardless of size, physical content, or mission carried out. Therefore it should be possible to develop a way of expressing these generic concepts for all enterprises in one form. This is the goal of an Enterprise Reference Architecture.

The architecture of an enterprise can be defined as a structural set of 'models' which represent invariant building blocks of the whole enterprise. The architecture can be considered as a basis for the design and implementation of the whole system of the enterprise. These models, which contain the invariant elements of the system and the relations between these elements, describe respectively the WHAT (what the enterprise system is conceptually composed of) and the HOW (how such a system technically works) and show how to transform the models into realities, i.e. the working system [9]. Thus the architecture is a structure or framework showing the interrelationship of a large number of 'different and separate' models describing parts of the system or their functions (i.e. all or part of the enterprise being considered).

A reference architecture is also a collection of the overall generic functions, descriptions, or behaviors of many (hopefully all) types of systems. This collection can be used for the overall description of these systems, and for the discussion of the overall characteristics, actions or needs of each such system. This document was called an Architecture to distinguish it from the already published Purdue CIM Reference Model [10]. Either term would apply to either document with the interpretation now given the word, architecture, in the CIM literature.

8.8 EXPANSION OF PERA TO COVER ANY TYPE OF ENTERPRISE (OTHER THAN MANUFACTURING OR PRODUCTION)

The preceding discussion leads to a set of principles concerning the expansion of this Architecture to include all types of enterprises in its capability to model the development of a program, project or system.

- All functions and operations on the right hand (Manufacturing) side of the Purdue Reference Architectural Diagram can be represented as relating only to Level 0 of the Purdue Scheduling and Control Hierarchy view of the CIM Reference Model [10]. (See Fig. 8.9, which illustrates the Purdue hierarchy diagram.)
- Conversely all functions and operations shown on the left hand (information) side of the Purdue reference architectural diagram relate only to Level 1 and higher of the Purdue Scheduling and Control Hierarchy view (CIM Reference Model [10]).
- All functions and operations on the right hand (Manufacturing) side can also be shown to relate only to services for the customer, i.e. operation and maintenance of the manufacturing facility to produce products for sale to customers.
- Again conversely all functions and operations on the left hand side (Informational) relate only to the wellbeing of the Business Entity itself, i.e. to its operational control in order to achieve the optimal operating conditions at hand.

The reader will undoubtedly have the most trouble in accepting the implications of Concept 4 above – that the left hand or information side is devoted solely to the wellbeing of the information and control system of the factory containing the integrated system. This is due to two reasons:

- In many cases the products of the factory include a considerable amount of data or information concerning the product and its composition, quality, etc., and often about the company as well.
- The latest 'buzz words' of the TQM (Total Quality Management) process state that, 'everybody works for the customer,' thus apparently violating Concept 4 or vice versa.

These can both be explained. Up to now we have been using the manufacturing system producing a simple product as a discussion example. Frankly, this was done to avoid Problem 2 above as long as possible in order to establish the architecture's principles in the mind of the reader before such questions appeared.

The easiest explanation is to say that, 'The whole company works for the customer.' Thus if the company operates to achieve the highest quality (a control and hence information function) and if the company operates to achieve minimum cost operation or maximum throughput as desired, (both control and hence information functions) then the company has achieved its maximum 'wellbeing' as noted. But the customer has also benefitted from the higher quality, lower cost product, or shorter delivery time possible. Thus these two concepts (TQM and the wellbeing of the enterprise) are not

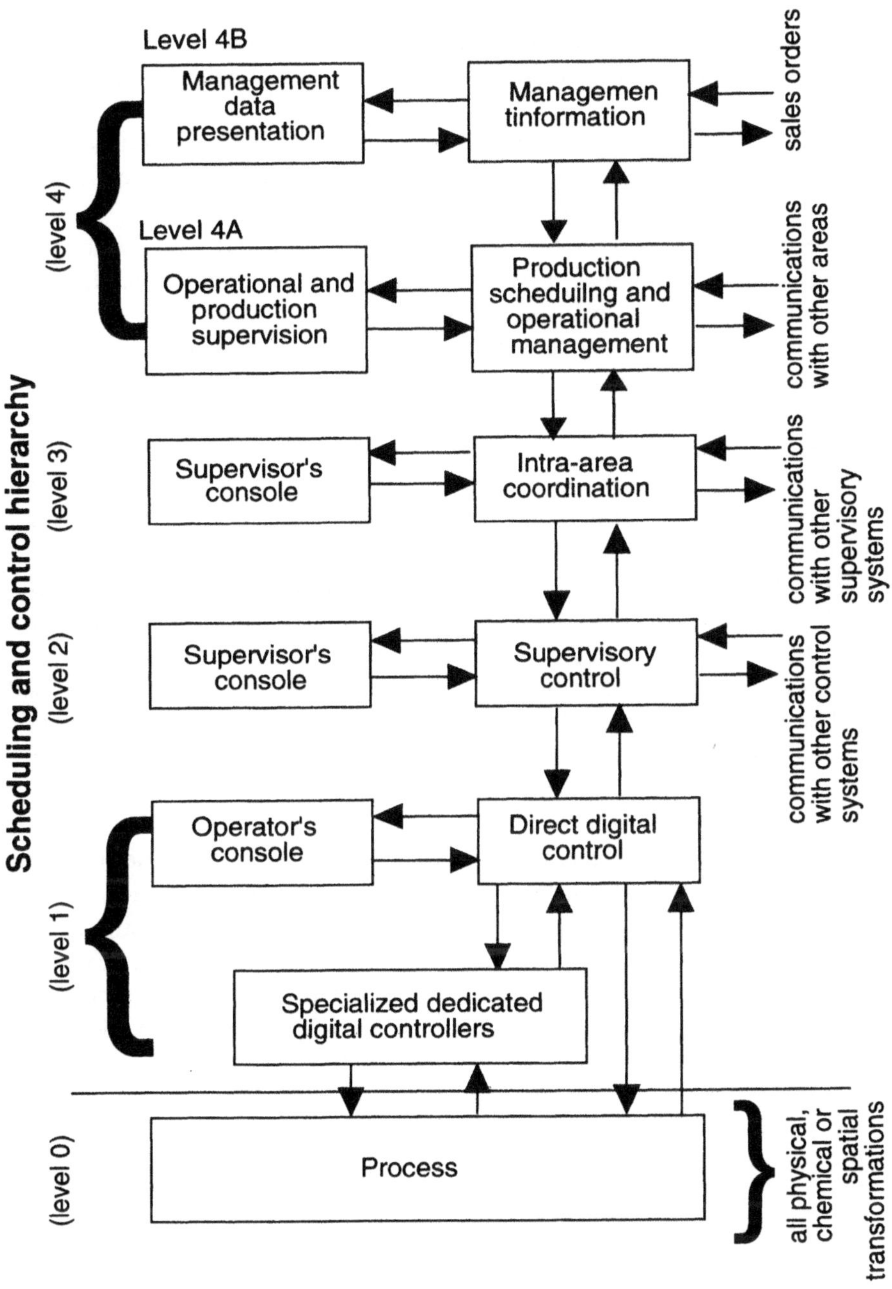

Figure 8.9 Assumed hierarchical computer control structure for an industrial plant (continuous process such as petrochemicals).

contradictory but complimentary. By being devoted to the wellbeing of the enterprise, the information system achieves the best service to the customers if the requirements are properly stated so that both goals are equally supportive.

The sale of information as a product is the major additional thrust of our discussion of the expansion of the applicability of the Architecture to cover any type of enterprise as will be seen from the material to follow.

Therefore those enterprises whose missions are strictly of an information services nature can also be treated in exactly the same manner as the manufacturing type enterprises just discussed.

As noted earlier, the right hand side of the Business Entity diagram relates to customer services while the left hand side relates to informational services to the enterprise itself. Thus the diagram can be expanded in its coverage to treat all enterprises – not just manufacturing or a information implementation program. Likewise the relationship to the Purdue Reference Hierarchy (Level 0 vs Levels 1–5) also holds true for the general enterprise. The next figures show this. (Figures 8.10 and 8.11).

The Enterprise is a collection of Business Entities (one or more), each of which performs a distinct service to its customers or those of the enterprise. It is often important to split up such an enterprise into separate Business Entities each carrying out its own individual distinct services by the methods being discussed here. Fig. 8.12 shows such a separation of a mission into its distinctly different component parts for a possible example of a manufacturing company. Note that the subunits considered need not be complete selfgoverned entities to qualify. They can be subdivisions of any size or governance. The criteria for separation here is whether or not their mission is sufficiently different from that of the parent organization or their fellows that such a separation in modelling would greatly simplify the modelling task. It is to be considered mainly as a technique for easing the task of a planning or design team in studying the enterprise's structure in relation to the capabilities of the Purdue Enterprise Reference Architecture.

Customer response or the provision of customer services may be carried out in many ways.

- By physical things (i.e. manufactured products; (the type of Business Entity discussed in all the earlier parts of this chapter).
- By pure physical services (transportation of goods or persons, availability of goods or service for purchase, rental (lease), etc., or
- Through the supply of information services (data), (information) to be used by others.

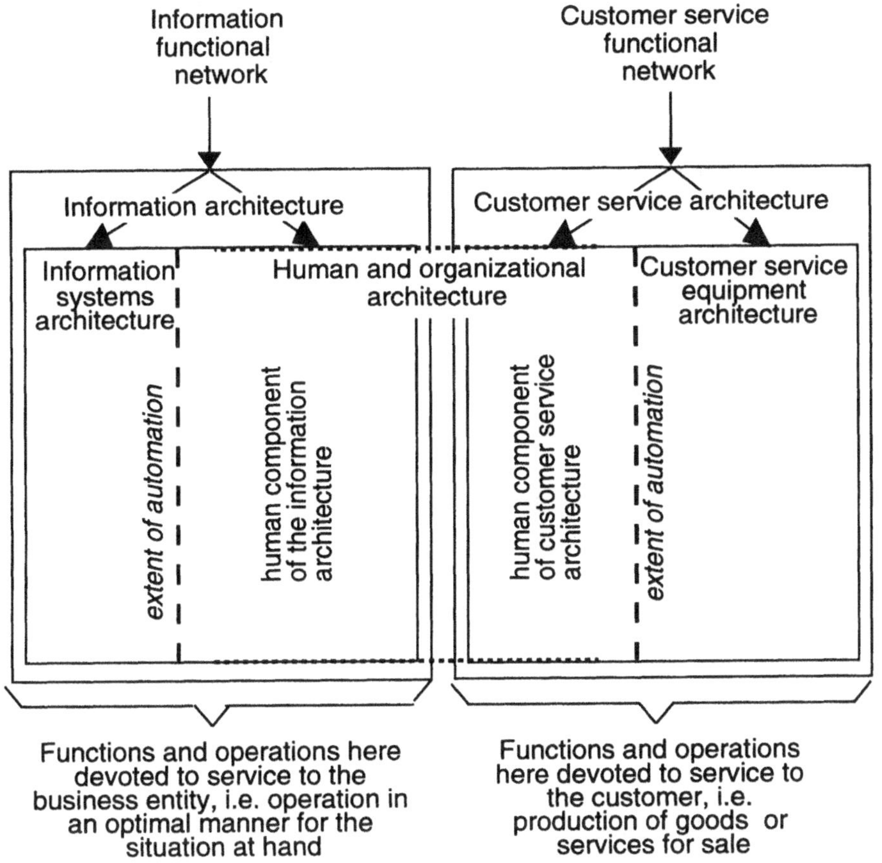

Figure 8.10 Further explanation of the definition of the generic enterprise by the Purdue Enterprise Reference Architecture.

Fig. 8.13 uses the information just developed to modify the labelling of the overall enterprise evolution diagram (Figures 8.8a and 8.8b) to cover the corresponding generic enterprise (i.e. the production of either goods or services or both). Fig. 8.14 shows how the definition of the tasks of the humans in each section of the Human and Organizational Architecture would change in order to be able to carry out the designated mission. For example, an information services company would employ information services type personnel to perform the desired customer service and thus would have these kinds of personnel in both compartments of the Human and Organizational Architecture. The reader should understand that Fig. 8.13

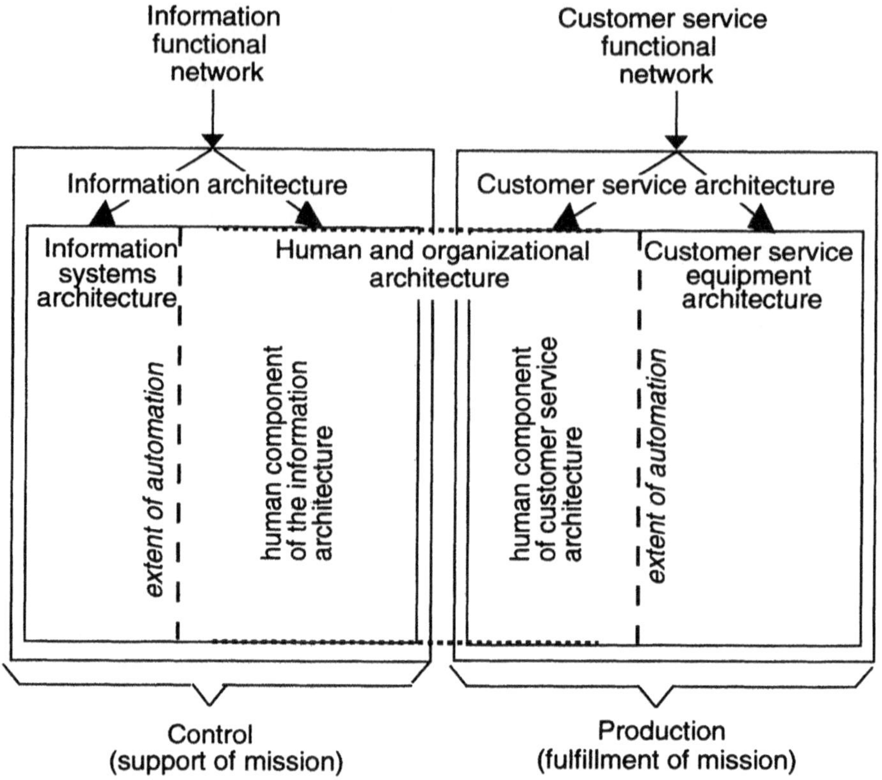

Figure 8.11 Further explanation of the definition of the generic enterprise by the Purdue Enterprise Reference Architecture.

could be extended to cover the total life history of that enterprise as discussed with Fig. 8.8a and 8.8b in the case of the manufacturing system. The notations there would bear the same general relationship as used in Figures 8.8a and 8.8b.

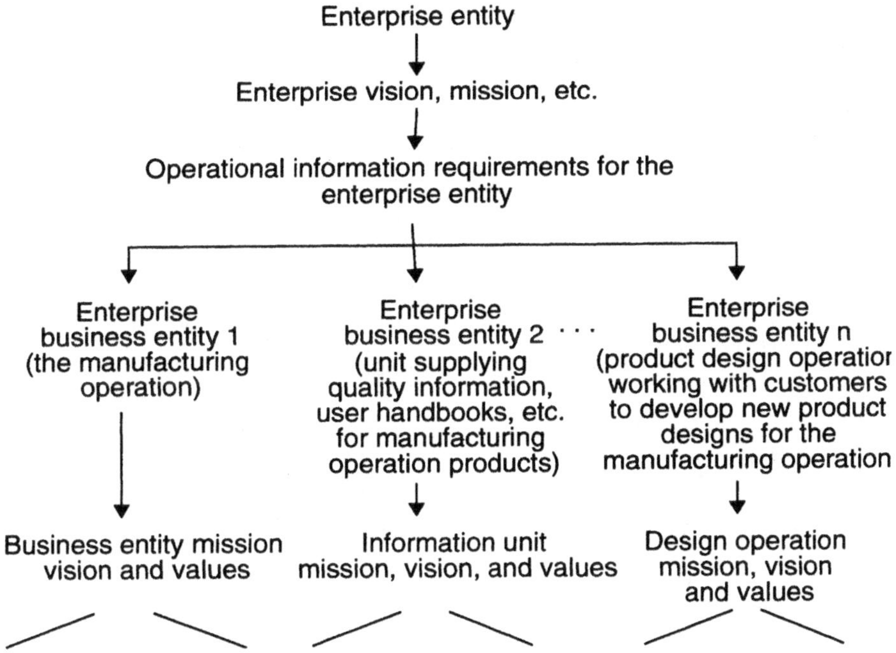

Figure 8.12 Subdivision of the overall enterprise into separate enterprise entities to properly model different enterprise activities (example of possible diversity of subdivision missions).

8.9 CIM SYSTEM IMPLEMENTATION CONSIDERATIONS AS SHOWN BY PERA AND THE PURDUE METHODOLOGY

8.9.1 How the Purdue Methodology works (the master plan is the key)

The Purdue Enterprise Reference Architecture [1] was originally developed to facilitate the preparation of The Implementation Procedures Manual for Developing Master Plans for Computer Integrated Manufacturing [11] by the Industry–Purdue University Consortium for CIM. The resulting Methodology can be stated as follows:

1. The methodology uses the Purdue Enterprise Reference Architecture [1] as the pattern and the framework for the overall program.
2. It uses the Purdue Reference Model for CIM [10] as the basis for much of the initial data and functional analysis information necessary.

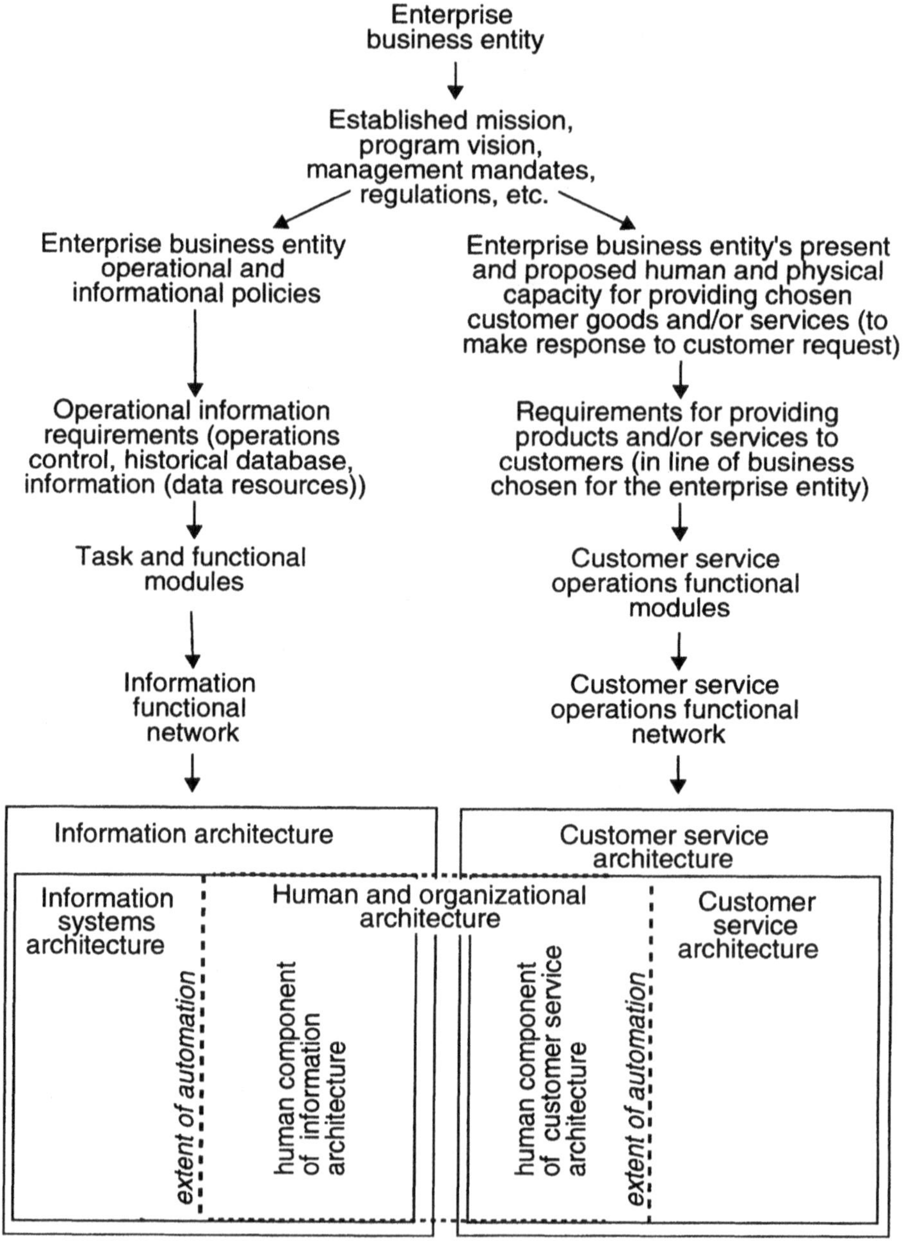

Figure 8.13 The overall diagram for the generic enterprise.

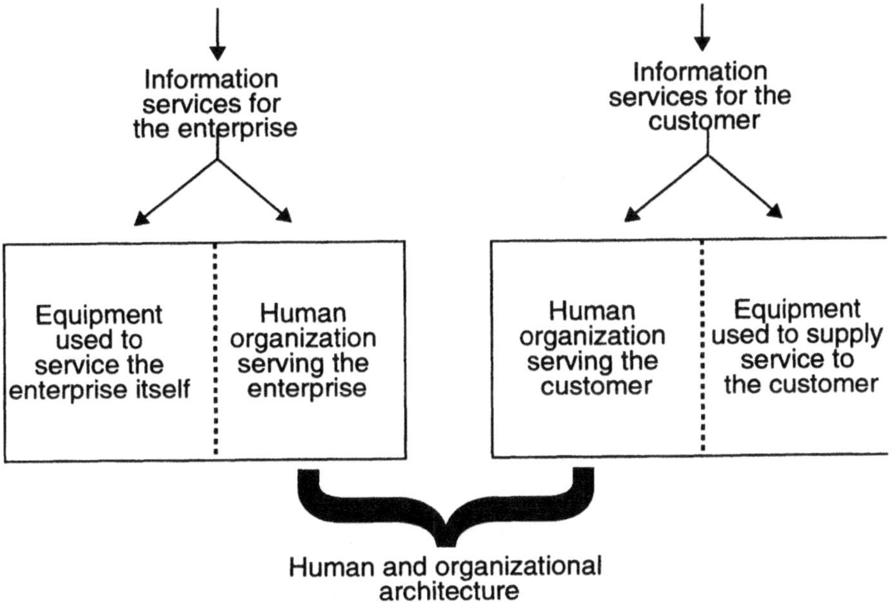

Figure 8.14 Services which are of an innovative information type will be different from the physical service type as shown in this diagram.

3. It uses the Implementation Procedures Manual [11] to help schedule and define the work to be done and to teach the how of master planning.
4. As part of the master planning effort, the methodology develops a CIM Program Proposal which is a set of projects, each within the resource capabilities of the enterprise in terms of manpower and capital, which taken together, fulfill the recommendations of the Master Plan.
5. These projects are then implemented by the Enterprise, as resources permit, following the recommendations of the Master Plan in each case, with the knowledge that on total completion the integration project will also be complete.

What is needed therefore, for each company contemplating a major integration effort, is for the company to develop a Master Plan covering all of the anticipated effort required to integrate the whole of the company or factory operation.

After this, smaller projects within the monetary and personnel resources capability of the company can be initiated with the knowledge that the sum of this and all succeeding projects will result in the final total integration of the company's activities. This will be possible provided that the requirements of the initial planning effort or Master Plan be followed in each and every one of the resulting projects.

But the detail and effort required for even the master planning activity is itself large and if done improperly will only lead to difficulties later. Thus there is a need for a methodology to assure that the Master Plan as developed by a company's CIM Planning Team is complete, accurate, properly oriented to future business developments and carried out with the minimum of resources (personnel and capital) necessary.

The Industry–Purdue University Consortium on An Implementation Procedures Manual for Developing Master Plans for Computer Integrated Manufacturing (CIM) has therefore developed such a methodology incorporated in the document of the same title [11]. The Manual presents a detailed description of the tasks involved in developing the Master Plan including its continual renewal. It gives the detail necessary both as to specifics and to the quantity of information and data needed. It specifies the interrelationship of the informational, the human and organizational and the physical manufacturing aspects of the integration considered; the management considerations and concerns; and the economic, cultural and technological factors involved.

8.10 USING THE IMPLEMENTATION PROCEDURES MANUAL

8.10.1 Development of the master plan using the implementation procedures manual

As noted earlier the Implementation Procedures Manual will provide the user group with all the guidance necessary to help them prepare the company's CIM Program Master Plan. This guidance includes the following information among others:

1. A set of activities that must be performed to develop the appropriate Master Plan for a specific plant site.
2. A detailed description of each activity including: resources required, input information required, methodologies or tools to be used, and the outputs or deliverables from each activity.
3. An approach that clearly ties the Master Plan to business needs.

Fig. 8.15 presents an overview IDEF0 diagram that illustrates how the Manual serves as the mechanism for the preparation of the Master Plan.

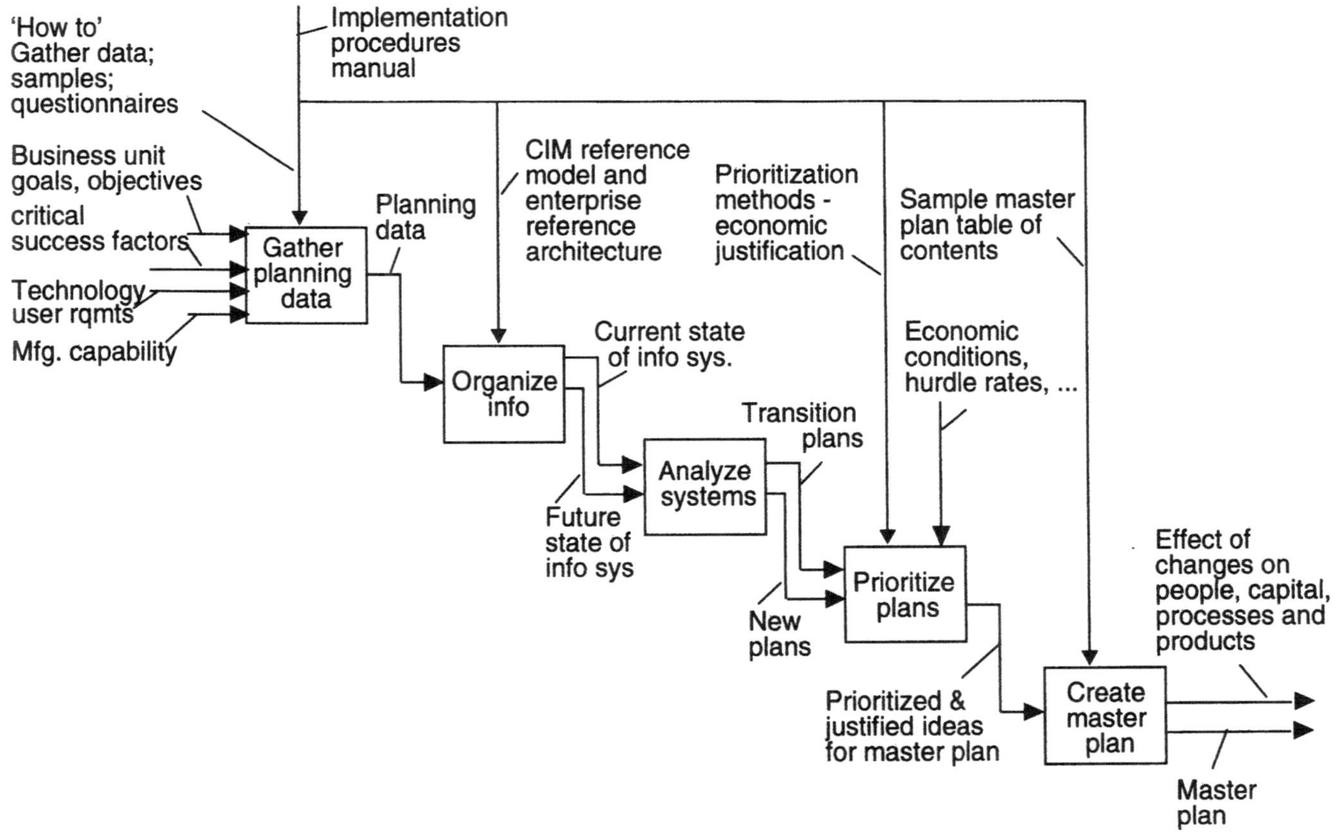

Figure 8.15 An overview of the preparation of a CIM Master Plan (context: analyse manufacturing).

8.10.2 Key elements in starting a CIM program

The Consortium Members in preparing this Implementation Procedures Manual believed that several key elements should be in place before any company should undertake an integration project. These key elements should be in place even before attempting the preparation of a Master Plan for an integration program. The first of these includes several very important individuals or groups of individuals.

These are: (1) a Champion, an individual, knowledgeable in CIM technology, who is anxious to promote it, and who serves as a catalyst to push towards such applications; and (2) Sponsorship, or support of upper management and user management personnel who, either separately or as influenced by the Champion, see the potential benefits of integration, and are willing to lend their prestige and influence to investigate and, if viable, promote the CIM program.

During the development of the Manual the Consortium members separated this sponsorship into two elements: (1) the Initiating Sponsor, a high level management individual who lends support and prestige to the work of the Champion and clears the hurdles, and (2) the Steering Committee, a group of stakeholders in the business unit for which the CIM Program is being developed, who lend direct management guidance and support to that effort.

The actual development of the Master Plan itself will be carried out by still another group, the CIM Planning Team. They do the actual analysis and preparation of the plan under the guidance of the Steering Committee.

The other element that must be in place before the CIM Program is officially initiated is the results of a Preliminary Economic Analysis that shows that the Company does have the potential for undertaking one or more potentially successful integration projects.

8.10.3 An outline of the implementation procedures manual

In order to carry out the purpose of simplifying the preparation of a Master Plan as much as possible the Manual consists of three main parts. They are: Planning for the Plan or getting ready to develop a Master Plan (Sections I and II); Developing the Master Plan (Section III); and Updating the Master Plan and Continuing Education (Section IV).

In addition there is a group of Appendices that supply additional information to the process of developing the Master Plan. Table 8.1 shows the

Table 8.1 Implementation procedures manual, showing organization into sections and blocks

Chapter	*Title*
Section I	Introduction
Chapter I-1	Management presentation summary
Chapter I-2	Introduction
Chapter I-3	Using the implementationprocedures manual
Section II	Preparations for the cim task
Chapter II-1	The task ahead
Chapter II-2	Get management support
Chapter II-3	Build consensus
Chapter II-4	Affirm the enterprise strategies
Chapter II-5	Identify and build the CIM planning team
Section III	Prepare the master plan
Block 1 (strategic – affirm strategies)	
Chapter III-1	Define the CIM business entity
Chapter III-2	Document the objectives, strategies, business plans,goals and critical success factors to be attained by CIM in the business entity
Chapter III-3	Affirm the to-be manufacturingpolicies
Block 2 (definition – define 'to-be')	

Table 8.1 Implementation procedures manual, showing organization into sections and blocks (Continued)

Chapter	*Title*
Chapter III-4	Identify significant initiatives and opportunities enabled by cim to achieve goals and objectives
Chapter III-5	Define to-be manufacturing functional architecture
Chapter III-6	Define to-be information functional architecture
Chapter III-7	Define to-be human and organizational functional architecture
Block 3 (definition - define 'as-is')	
Chapter III-8	Document the as-is manufacturing functional architecture
Chapter III-9	Document the as-is information functional architecture
Chapter III-10	Document as-is human and organizational functional architecture
Block 4 (planning – transition planning)	
Chapter III-11	Identify required standards selection process
Chapter III-12	Identify logical transition path from as-is to to-be
Chapter III-13	develop training plan
Chapter III-14	Identify feasible solutions in the form of a set of cim projects
Chapter III-15	Analyze costs, benefits and risks for proposed projects
Block 5 (Strategic – finalize master plan)	

Table 8.1 Implementation procedures manual, showing organization into sections and blocks (Continued)

Chapter	*Title*
Chapter III-16	Final critical evaluation of the master plan contents
Chapter III-17	Author master plan and deliver for action
Section IV	Process of renewal
Chapter IV-1	Master plan renewal process
Chapter IV-2	Plan continuing education and training
Appendix I	Modelling methods and tools
Appendix II	CIM business entity database
Appendix III	Master plan definitions andexamples
Appendix IV	Benchmarking (share and share alike)
Appendix V	Glossary
Appendix VI	A bibliography of the technical literature relative to the task of the implementation procedures manual
Appendix VII	A list of individuals who contributed to this implementation procedures manual

chapter title list for the Manual indicating their further collection into related subject blocks under the major sections as just noted above.

Fig. 8.16 shows still another method of illustrating the subject matter flow through the Manual and the suggested Master Plan format. Fig. 8.17 outlines the correspondence of the coverage of the Enterprise Reference Architecture and of the CIM Reference Model in relation to the factory, and to the overall enterprise and its business environment.

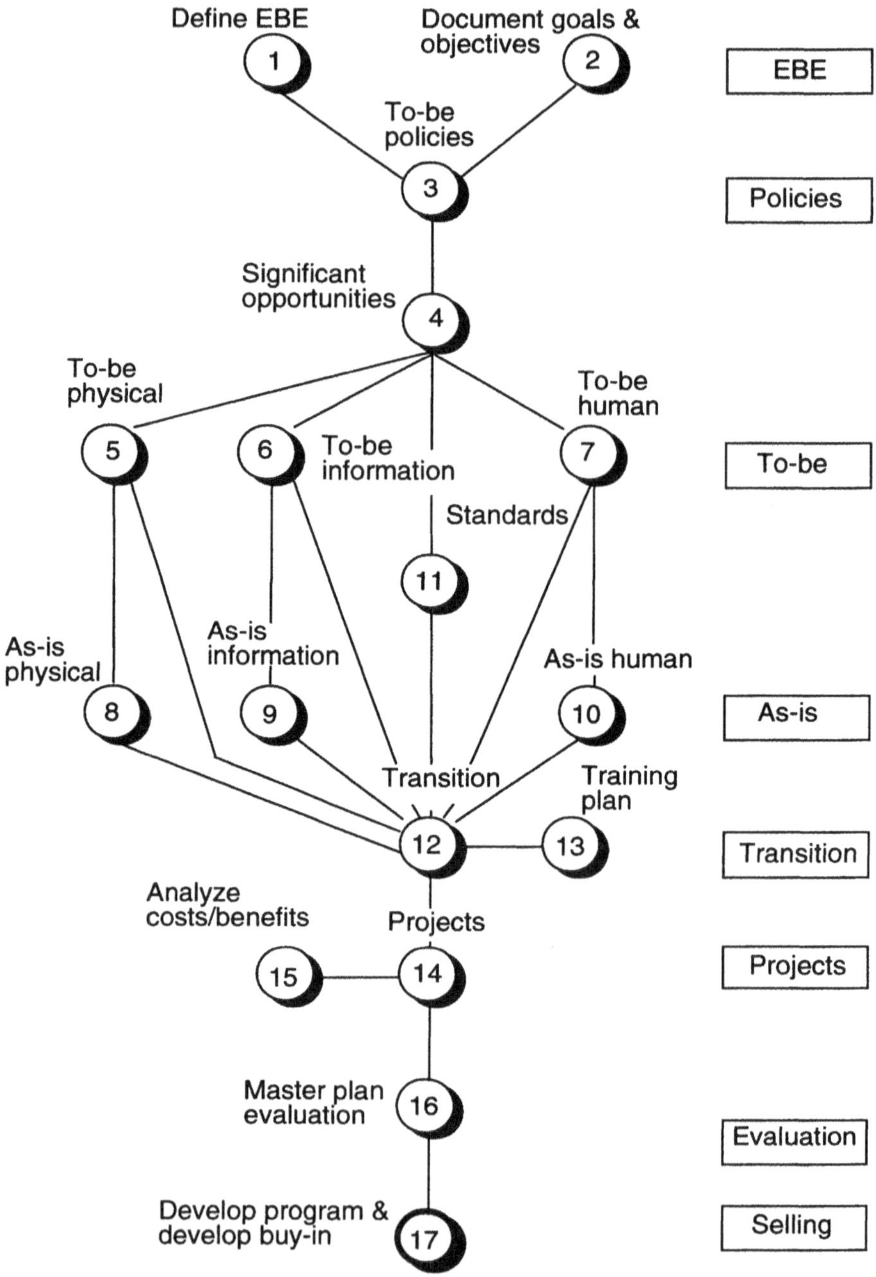

Figure 8.16 Purdue CIM Implementation Procedures Manual flow (numerals refer to manual chapter numbers).

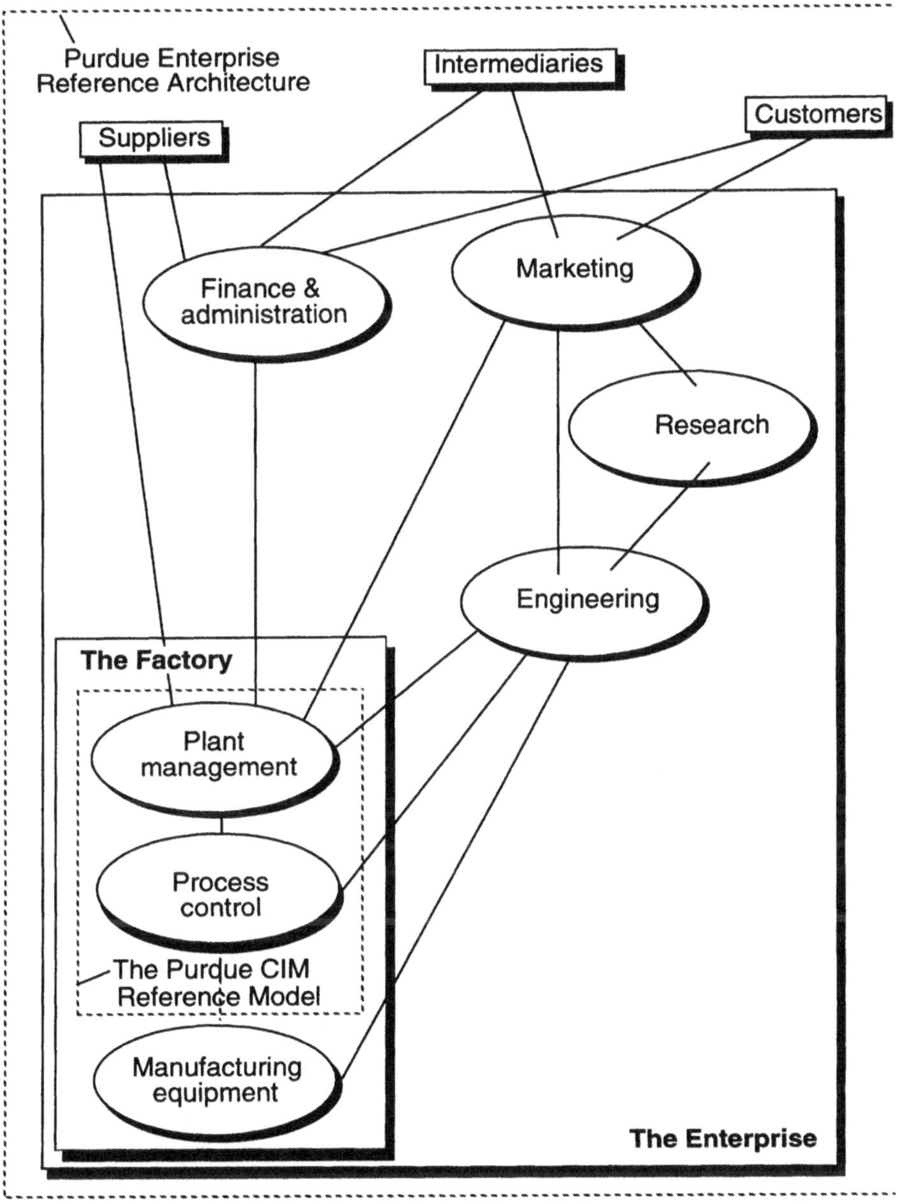

Figure 8.17 Purdue methodology tools in relation to the overall manufacturing enterprise.

8.11 REFERENCES

[4] Anonymous, IDEF0 Function Modeling Manual, ICAM Program Office, Wright–Patterson AFB, OH (1981).
[6] Chen, Peter, 'Principles of Database Design,' In Database Design Based on Entity and Relationship, S. Bing Yao, Editor, Chapter 5, Prentice–Hall Inc., Englewood Cliffs, NJ (1985).
[2] DeMarco, T., Structured Analysis and System Specification, Yourdon Press – Prentice Hall, Englewood Cliffs, NJ (1979).
[9] Doumeingts, G., Vallepir, B., Zanettin, M., and Chen, D., GIM, GRAI Integrated Methodology, A Methodology for Designing CIM Systems, Version 1.0, Special Report, GRAI Laboratory, University Bordeaux I, Bordeaux, France (May 1992).
[8] ESPRIT Consortium AMICE, Open System Architecture, CIMOSA, AD 1.0, Architecture Description, ESPRIT Consortium AMICE, Brussels, Belgium (1991).
[11] Industry–University Consortium, Purdue Laboratory for Applied Industrial Control, An Implementation Procedures Manual for Developing Master Plans for Computer Integrated Manufacturing, Report Number 155, Purdue Laboratory for Applied Industrial Control, Purdue University, West Lafayette, IN (June 1992).
[5] Ross, D. T., 'Applications and Extensions of SADT,' IEEE Computer, Vol 18, No. 4, pp 25–34 (April 1985).
[7] U.S. Air Force (no author) 'Integrated Computer-Aided Manufacturing (ICAM)', Architecture Part II, Vol. V –Information Modeling Manual (IDEF1), AFWAL-TR-81-4023, Wright–Patterson Air Force Base, OH, (June 1981).
[10] Williams, T. J., Editor, A Reference Model for Computer Integrated Manufacturing (CIM), A Description From the Viewpoint of Industrial Automation, Minutes, CIM Reference Model Committee, International Purdue Workshop onIndustrial Computer Systems, Purdue University, West Lafayette, IN (1988), Instrument Society of America, Research Triangle Park, NC (1989).
[1] Williams, T. J., The Purdue Enterprise Reference Architecture, Instrument Society of America, Research Triangle Park, NC (1992).
[13] Yourdon, Edward, and Constantine, L.L., Structured Design: Fundamentals of a Discipline of Computer Program and Systems Design – Edition 2, Prentice–Hall, Englewood Cliffs, NJ (1979)

8.12 FURTHER READING

Anonymous, Data Processing – Open Systems Interconnection – Basic Reference Model, ISO DIS 7498, International Standards Organization, Geneva, Switzerland (December 3, 1980).

9

Introduction to the evaluation methods used

P. Bernus

9.1 METHOD 1 – QUESTIONNAIRES

Once the Task Force had decided that the major elements of its technical charge involved an evaluation of how well the three candidate architectures (CIMOSA, GRAI-GIM and Purdue) fulfilled the needs for 'integrating manufacturing activities and enterprises', the major task facing them was to choose the method(s) of evaluation to be used.

An extensive questionnaire was developed to cover all important ramifications of the requirements for this architectures. This would then be completed by the groups developing each separate architecture and reviewed by the Task Force before being adopted for this report. After several reviews Version Number 3 was adopted and used. Copies of the Version 3 Questionnaire itself and the corresponding answers for each of the candidate architectures are presented here in the next Chapter.

Since the Questionnaire is quite extensive, it was suggested that a shorter questionnaire be developed and used. This suggestion was accepted and the attached Short Form Evaluation Questionnaire for Integration Reference Models was prepared and completed for all three candidate architectures. They are presented at the end of the next Chapter.

9.2 METHOD 2 – MAPPINGS, ONE ON ONE

One possibility to find how the candidate architectures correspond to one another is to 'map' the characteristics and capabilities of each of the architectures onto a suitable representation of the corresponding characteristics and capabilities of the others (a total of six, hopefully graphical presentations, to cover the full spectrum of characteristics and capabilities). To be able to do this it was proposed that a diagram representing the life-cycle for each architecture be used for the framework desired. These would be suitably modified to include any capabilities missing in that particular architecture if necessary. Representations of CIMOSA on Purdue and vice versa were developed first, followed by the related GRAI-GIM representations (these are recorded here in Chapters 11–13).

9.3 METHOD 3 – MAPPING AGAINST A SET OF REQUIREMENTS

A second possibility to find the correspondances is to develop a different form of a mapping diagram, based on a Matrix of Requirements or Needs. The first version of this was developed as of July 1992 and critiqued by correspondence at subsequent Meetings. The final form was then given a set of trial mappings for each architecture. These latter were then presented to representatives of CIMOSA and GRAI-GIM during December 1992.

The results are included in this report in Chapter 17.

10

Evaluation questionnaires

P. Bernus and L. Nemes

As noted in Chapter IV-1, the Evaluation Questionnaire method of reviewing the characteristics and capabilities of each of the candidate architectures was proposed and developed. A copy of the Questionnaire[1] itself plus completed copies prepared by the development groups or in consultation with the developers of each of the architectures is presented here.

These are followed by a commentary on the results giving an interpretation of the capabilities and characteristics of these architectures based on the responses received.

Because of the length and complexity of the Questionnaire a simple one-sheet form is presented at the end of in this Chapter along with the corresponding responses prepared by each of the development groups.

1. The questionnaire was refined several times by the Task Force. The present version is version 3

10.1 ARCHITECTURE

10.1.1 Users of the architecture

Who is the intended/targeted audience of this architecture?

10.1.2 Intended scope of architecture

Please check those activities which are in the intended scope of the architecture (in the short term, i.e. 2–3 years).

[] covered context	[] partially covered context	[] context not covered	[] no information

Example

[ ] [ ] [ ] [ ] product research and development
[ ] [ ] [ ] [ ] customer services, product maintenance
[ ] [ ] [ ] [ ] product management, marketing function, strategic
[ ] [ ] [ ] [ ] product distribution
[ ] [ ] [ ] [ ] marketing
[ ] [ ] [ ] [ ] product design
[ ] [ ] [ ] [ ] production planning
[ ] [ ] [ ] [ ] manufacturing control
[ ] [ ] [ ] [ ] manufacturing
[ ] [ ] [ ] [ ] material handling, storage and transport
[ ] [ ] [ ] [ ] testing and quality control
[ ] [ ] [ ] [ ] logistics
[ ] [ ] [ ] [ ] acquisition
[ ] [ ] [ ] [ ] legal services
[ ] [ ] [ ] [ ] personnel services finance, accounting
[ ] [ ] [ ] [ ] infrastructure, archiving and library services
[ ] [ ] [ ] [ ] infrastructure, communications (excluding computer
[ ] [ ] [ ] [ ] communications)
[ ] [ ] [ ] [ ] information technology infrastructure, including computer
[ ] [ ] [ ] [ ] communications, computing, database management, etc.
[ ] [ ] [ ] [ ] education
[ ] [ ] [ ] [ ] strategic enterprise management
[ ] [ ] [ ] [ ] market research
[ ] [ ] [ ] [ ] resource management
[ ] [ ] [ ] [ ] factory development /development of the production facility)
[ ] [ ] [ ] [ ] project management
[ ] [ ] [ ] [ ] factory design

[ ][ ][ ][ ] factory building and modernization
[ ][ ][ ][ ] factory maintenance
[ ][ ][ ][ ] infrastructure, energy
[ ][ ][ ][ ] infrastructure, buildings and grounds
[ ][ ][ ][ ] other (please specify)

10.1.3 What is the highest intended level of genericity on which this architecture is applicable? (check one)

[ ] every type of enterprise
[ ] certain type of industry (please specify which one)?
[ ] certain group or type of enterprises(please specify which ones)?
[ ] certain type of technology or product (please specify which one)?
[ ] certain type of function (please specify which one)?

What is the area of industry in which the architecture has already been applied?

In what area of industry are there current plans to apply the architecture?

10.1.4 Which stage of systems development is addressed by the architecture? (multiple choice question)

[] fully covered	[] partially covered	[] not covered	[] no information

Example

[ ][ ][ ][ ] requirements
[ ][ ][ ][ ] implementation (configuration and reconfiguration)
[ ][ ][ ][ ] design
[ ][ ][ ][ ] operation and maintenance

10.1.5 What are the goals which we would like to attain by developing generic CIM architectures? (Which are the main factors of industry's need?)

(Please concentrate on the user requirements of a prospective utilizing industry.)

[] essential factor	[] important factor	[] advantageos but not essential	[] not important

Example

[ ][ ][ ][ ] better use of current resources
[ ][ ][ ][ ] integration of current technology

[ ] [ ] [ ] [ ] cutting excessive costs of having to develop CIM systems indi-individually
[ ] [ ] [ ] [ ] improve the quality of the developed individual systems/ manufacturing systems/CIM systems,etc.
[ ] [ ] [ ] [ ] enabling the development of a marketplace for compatible CIM products
[ ] [ ] [ ] [ ] decrease response/product cycle time
[ ] [ ] [ ] [ ] contain metrics for exploring economic/technological options
[ ] [ ] [ ] [ ] decisional integration
[ ] [ ] [ ] [ ] understanding and training
[ ] [ ] [ ] [ ] economic aspects
[ ] [ ] [ ] [ ] obtain flexibility in system and adaptation to change in requirements
[ ] [ ] [ ] [ ] Other (please specify)

10.1.6 De facto scope and detail

This question is about how completely the architecture has been defined relative to its intended scope (see Question 2) and its intended detail.

You will have by now a few projects which have already used one part or another of this architecture. We would like to have a list of those activities which were covered by these projects.

To complete the questionnarie please:

(a) Make a copy of the sheets on which you answered Question 2. Make copies for each different project you intend to describe.

(b) Indicate, with a cross, X, on the copies you have made, the activities covered by this project.

(c) Please characterize the maturity of the models produced.

Circle the corresponding statement if it refers to models which have repeatedly been used in industry; underline the corresponding statement if it marks a feasibility tested but not yet industry mature model; provide, without any mark, those statements which refer to models under development. Example (you should end up with annotated copies of Question 2 like this):

[ ] [ ] [ ] [ ] marketing
[x] [ ] [ ] [ ] product design: *product design database*, *turbine blade*
[ ] [ ] [ ] [ ] production planning
[ ] [ ] [ ] [ ] manufacturing control
[ ] [ ] [ ] [ ] manufacturing
[x] [ ] [ ] [ ] material handling, storage and transport: *AGV functions*
[ ] [ ] [ ] [ ] testing and quality control
[ ] [ ] [ ] [ ] logistics

The above example shows that:

- a model for product design databases has been feasibility tested
- a model for turbine blade design is under development, and
- a model for automatically guided vehicles (AGVs) has been developed and feasibility tested

10.1.7 Omissions

(a) Please check and give details if needed:

[ ] the architecture covers its promised intended scope
[ ] the architecture omits some necessary detail (please specify the additional detail the architecture should provide)

(b) If you indicated a need for extension, could you name here any other architecture and model which does cover the missing detail (give specifics)

10.1.8 Redundancy

Does this architecture require that existing standards must be rewritten (e.g. expressed in a new form without changing the content) in order to fit this architecture?

10.1.9 System extensibility

How do you ensure that systems created according to this architecture are open ended?

If you judged that the architecture allows modular evolution, is this judgement based on

[ ] theoretical considerations
[ ] feasibility study or pilot project
[ ] repeated industry experience
[ ] other (please sepcify):

10.1.10 Resilience

Does the architecture provide any principles or guidance that will help the user to limit the effect of imposed external changes on the operation of the system?

Example for resilience of an architecture with respect to distribution of computing power: if the communication between software modules is designed to be technology independent then a change from networked personal computers to distributed computing environment will not affect every single software module (only localized modification will be necessary).

Example for lack of resilience with respect to the underlying database technology: if the design database is based on the Codasyl data model and the access to it on the Codasyl data manipulation language then the change to relational or object oriented database management systems will profoundly affect every software module which uses the design database.

10.1.11 Skills required to implement

Does the architecture adequately treat the skills required by the user to allow for the design and building of a specific system described by it? (give rating):

[ ] poor (can you name the main reason why?)
[ ] adequate
[ ] good
[ ] very good (can you name the main reason why?)
[ ] no information
[ ] issue not addressed

The architecture is very good from this respect if the skills required are available, or there are tried and natural, pragmatic ways for practitioners to extend their skills in the new direction.

10.1.12 Dependence on new technologies

Does the implementation of this architecture require a given type of new technology for any application within the scope? (A technology is new in this respect if it has not yet been proven under industrial circumstances)

[ ] no new technology required
[ ] yes (please specify what new technology is involved) e.g.
[ ] distributed relational databases
[ ] distributed object oriented databases
[ ] OSI communication
[ ] Open Distributed Processing Platform
[ ] others (please specify)

10.1.13 What is your opinion about the future prospects for this architecture?

[] yes, widespread	[] yes, limited area	[] no	[] undecided	Example

[ ][ ][ ][ ] will have theoretical impact
[ ][ ][ ][ ] will gain industry acceptance in the present form
[ ][ ][ ][ ] will gain industry acceptance in developed (or modified) form
[ ][ ][ ][ ] other (please specify)

10.1.14 What were the reasons for the development of this architecture?

[] main reason	[] cofactor	[] not a known reason	[] no information	Example

[ ][ ][ ][ ] utilization in a given segment of industry
[ ][ ][ ][ ] development of a national standard
[ ][ ][ ][ ] development of an international standard
[ ][ ][ ][ ] utilization by a given firm
[ ][ ][ ][ ] utilization by a group of firms
[ ][ ][ ][ ] utilization for teaching purposes
[ ][ ][ ][ ] utilization for research purposes (Other, please specify)

10.1.15 What is the driving force behind this architecture?

[] main reason	[] cofactor	[] not a known reason	[] no information	Example

[ ][ ][ ][ ] technology pull (there was a perceived need in industry for which the generic architecture intends to be the answer
[ ][ ][ ][ ] technology push (results in the area of architectures methodologies, information technology, communications manufacturing technology created opportunity or incentive for for industry to develop this generic CIM architecture

Please expand your views on the above quesion if possible.

10.1.16 Adequacy

Please rate the adequacy of the architecture with respect to the help which it provides for attaining the essential or important goals which you identified in Question 5.

poor	average	good	excellent	no info.	Example
[]	[]	[]	[]	[]	

[ ] [ ] [ ] [ ] [ ] better use of current resources
[ ] [ ] [ ] [ ] [ ] integration of current technology
[ ] [ ] [ ] [ ] [ ] cutting excessive costs of having to develop CIM systems individually
[ ] [ ] [ ] [ ] [ ] improve the quality of the developed individual systems/manufacturing systems/CIM systems etc.
[ ] [ ] [ ] [ ] [ ] enabling the development of a marketplace for compatible CIM products
[ ] [ ] [ ] [ ] [ ] decrease response/product cycle time
[ ] [ ] [ ] [ ] [ ] contain metrics for exploring economic/technological options
[ ] [ ] [ ] [ ] [ ] decisional integration
[ ] [ ] [ ] [ ] [ ] understanding and training
[ ] [ ] [ ] [ ] [ ] economic aspects
[ ] [ ] [ ] [ ] [ ] obtain flexibility in system configuration and adaptation to change in requirements
[ ] [ ] [ ] [ ] [ ] other (please specify)

What is the main strength of this architecture?

To your knowledge which essential or important goals (see Question 5) are specifically not adequately addressed by this architecture?

10.2 MODELLING

10.2.1 What kinds of models describe this architecture?

Please characterize the models which this architecture requires to be produced? (The table for your answers has been broken into four parts because it had too many columns – see following pages).

Characterize the corresponding modelling languages used to express the models. NB. For the purposes of collecting information we have ordered these tables by the models and not by the modelling languages. Because the same modelling language is often used for more than one model, it is enough to characterize the language only once. However, it is expected that the questions about the language's adequacy for each separate type of model will be answered separately.

Important: the descriptions capturing requirements, designs, etc, are produced in many, very different forms (formal mathematical models, graphs, natural language, tables etc.) We use the word 'model' to refer to all of these forms.

Model[a]	Modelling Language[b]	Degree of formality[c]	Should it be executable?	Is it executable?	Demonstrated feasibility[d]

a. Please add rows as necessary. E.g., 'functional requirements model,' 'organizational design model', etc. Note that if models are substantially different on various levels of genericity then you are asked to enter these as separate models.
b. Here 'language' is used in general sense, e.g., ER, Lotos, IDEF0, natural language, Petri net, etc.
c. Please rate: I = Informal, S = Structured but the expressive power of the formal part is not adequate (missing information being added in natural language), F = Formal mathematical model, with formal semantics
d. This is the feasibility of producing an adequate model using this language, P = Practically demonstrated, T = Demonstrated in theory, or only C = Conjectured? Please add further comments in footnotes and on separate pages if necessary.

Model[a]	Adequacy[b] of modelling language	Is there any related[c] standard modelling language?

a. Please add rows as necessary.
b. Please rate P = Poor, A = Adequate, G = Good, V = Very good; please comment on problems, limitations in footnote if needed. E.g., overformalized, not formal enough, certain detail not expressible, forces unnecessary decision to be taken too early, etc.
c. If the language used is not a standard and there is a related standard, could you comment, in a footnote, why the selected language is used instead of a standard one?

Model[a]	Verifiability[b] of model	Could any alternative language be used?	Can completeness be achieved?[c]	Supported by CASE tool?[d]

a. Please add rows as necessary.
b. Rate: VH = model verifiable by human expert, following a method; VA =automatic or interactive verification is possible; VK = automatic or interactive verification is possible if large knowledge base present; VE = validation possible only by experiments (real or simulation).
c. Y = yes, a complete mathematical model can be developed from which all the necessary design features can be calculated, N = no, – if not, could you tell in a footnote what provision can be made to substitute for the lack of mathematical completeness?
d. Please provide names of CASE tools (or write Y=yes if name is unknown to you), N=no, U=unknown (no information). Note that question 20 will separately ask about the adequacy of the CASE/CAE tools listed.

Model[a]	Adequacy of language for people who should use/interpret the model[b]	Adequacy of language for those who should produce the model[c]

a. Please add rows as necessary.
b. Please rate the modelling language P = poor, A = adequate, G = good, V = very good; please comment on problems, limitations in footnote if needed
c. Please rate the modelling language P = poor, A = adequate, G = good, V = very good; please comment on problems, limitations in footnote if needed.

10.2.2 How well does the architecture describe the operational characteristics of an enterprise in contrast with describing only its basic design feature?

[ ] operational characteristics can be explicitly specified
[ ] operational characteristics can be derived from the design features (e.g. through simulation model)
[ ] operational characteristics not covered by this architecture
[ ] no information

10.2.3 Formalism

Do you think formal methods (with rigorously defined syntax and semantics) are needed to describe and interpret the architecture under discussion and the representation of particular systems by it?

[ ] necessary
[ ] desirable
[ ] possible
[ ] not possible

Please comment on what areas you have applied or intend to apply formal methods:
How well can the architecture be used to verify the completeness and adequacy of the overall design for an enterprise or any of its components? (Please comment on merits and problems.)
Could there be a formal canonical view developed together with a modelling language which would subsume all of the ones used? If yes, please specify what should be the characteristics of such a language:
Do you think that the scope of the formal methods is broad enough to capture what you want to describe?

[ ] yes
[ ] no
[ ] question not applicable
[ ] no sufficient information

Can people who need to deal with these models easily relate to these formalisms?

[ ] yes
[ ] yes, depending on the field/level
[ ] no
[ ] no, depending on the field/level

[ ] not applicable
[ ] not sufficient information

10.2.4 Verification support

How extensively are Computer Aided Engineering (CAE) tools available to verify the prescribed models?

[ ] full support for the following models:
[ ] some support for the following models:
[ ] no support for the following models:
[ ] not applicable
[ ] no sufficient information

Do you think it is necessary to have CAE tools to support the methodology?

[ ] yes, full support is needed
[ ] some support is needed
[ ] not needed
[ ] not applicable
[ ] no sufficient information

Are there any limitations to achieving the above?

10.2.5 Economic questions

How well does the associated methodology handle economic questions related to the enterprise described using its methods? (Give comments)

[ ] well covered
[ ] some, but not full coverage of issues (what is lacking?)
[ ] not addressed in the modelling
[ ] no information

10.3 METHODOLOGY

10.3.1 Does the architecture have a methodology connected to it?

[ ] no
[ ] yes (please give the name of the methodology)

If possible, can you specify what are the components of the methodology?

How complete is the methodology?

[ ] this is a complete methodology, covering all aspects of enterprise engineering activity
[ ] the methodology is a collection of methods/logical techniques (but only separate methodologies are available for the individual modelling tasks)
[ ] the methodology is planned to be a complete one

Does the methodology address the human factor in the enterprise?

[ ] treats the necessary human organization
[ ] skill levels for operating personnel
[ ] training requirements
[ ] treats teams
[ ] treats individual workers

10.3.2 Validation status of the methodology

Was the methodology validated

[ ] in implemented industry projects/systems
[ ] in experimental systems
[ ] other: (Could you give references?)

10.3.3 Maturity of the methodology

The methodology has been

[ ] successfully taught by others than its developers
[ ] successfully applied by other than its developers
[ ] repeatedly applied by its developers
[ ] shown to work by its developers
[ ] other (e.g. 'under development')

10.3.4 Scope of the methodology

[] extensive coverage	[] some coverage	[] no coverage	[] no information	[] not applicable	Example

Check appropriate fields of the table. Make sure you understand how to read together the names of rows and columns to make sense of answer boxed! (An example is given below the table.)

Activities and related methods to ↙	...common to all systems in the scope	...common to a family of systems in the scope (which one?)	...of individual systems
produce requirements...	[ ][ ][ ][ ][ ]	[ ][ ][ ][ ][ ]	[ ][ ][ ][ ][ ]
produce the design specifications...	[ ][ ][ ][ ][ ]	[ ][ ][ ][ ][ ]	[ ][ ][ ][ ][ ]
produce the detailed design...	[ ][ ][ ][ ][ ]	[ ][ ][ ][ ][ ]	[ ][ ][ ][ ][ ]
produce the system...	[ ][ ][ ][ ][ ]	[ ][ ][ ][ ][ ]	[ ][ ][ ][ ][ ]
produce the operational and maintenance methods....	[ ][ ][ ][ ][ ]	[ ][ ][ ][ ][ ]	[ ][ ][ ][ ][ ]

Example: check 'extensive coverage' box in upper right hand corner if the methodology extensively tackles 'activities and related methods to produce the requirements of individual systems,' to be read from the table as:

Activities and related
methods to ↘
produce the requirements ... ...of individual systems.

Note 1: If the methodology is partially complete in any of these respects. Please add footnote to the box you checked.

Note 2: You can add additional model types below, or on a separate sheet, if you feel that the above categories are not exhaustive:

10.3.5 Presentation of the methodology

[] extensive planned	[] some planned	[] extensive now	[] some now	[] none	[] no info.	[] not applicable

[ ][ ][ ][ ][ ][ ][ ] handbook or collection of handbooks
[ ][ ][ ][ ][ ][ ][ ] teaching material
[ ][ ][ ][ ][ ][ ][ ] organizational procedures
[ ][ ][ ][ ][ ][ ][ ] CASE tool support for individuals
[ ][ ][ ][ ][ ][ ][ ] project database
[ ][ ][ ][ ][ ][ ][ ] CASE tool support for project team
[ ][ ][ ][ ][ ][ ][ ] any other form? (please specify):

10.3.6 Genericity

We want to know how big the gap is between the methodology as presented (see Question 26) and the way it is applied in particular projects.

[ ] the methodology describes general procedures only
[ ] the methodology is supported by a collection of application examples / cookbook

10.3.7 Quality of the methodology

Please rate the methodology from the following aspects:

poor []	average []	good []	excellent []	no info. []	N/A []

Example

[ ][ ][ ][ ][ ][ ] support of cooperative teamwork
[ ][ ][ ][ ][ ][ ] support of work with multiple organizations
[ ][ ][ ][ ][ ][ ] flexibility in terms of project size
[ ][ ][ ][ ][ ][ ] flexibility in terms of innovativity of project
[ ][ ][ ][ ][ ][ ] quality control of project
[ ][ ][ ][ ][ ][ ] CAE tools support

11

Evaluation questionnaires for CIMOSA

Task Force

11.1 ARCHITECTURE [1] (All answers in italics)

11.1.1 Users of the architecture

What is the intended/targeted audience of this architecture?

No answer provided

11.1.2 Intended scope of architecture

Please check those activities which are in the intended scope of the architecture (in the short term, i.e. 2–3 years)

[] covered context	[] partially covered context	[] context not covered	[] no information	Example

[x] [ ] [ ] [ ] product research and development
[ ] [ ] [ ] [x] customer services, product maintenance
[ ] [ ] [ ] [x] product management, marketing function, strategic
[x] [ ] [ ] [ ] product distribution
[ ] [ ] [ ] [x] marketing
[x] [ ] [ ] [ ] product design
[x] [ ] [ ] [ ] production planning
[x] [ ] [ ] [ ] manufacturing control
[x] [ ] [ ] [ ] manufacturing
[x] [ ] [ ] [ ] material handling, storage and transport

1. Questionnaire filled in by Peter Bernus based on Questionnaire Version 2; answers provided by Ing. Jakob Vlietstra, and information provided at the Melbourne Task Force Meeting by Ing. Dick Zoetekouw (May 1992)

[x] [ ] [ ] [ ] testing and quality control
[x] [ ] [ ] [ ] logistics
[ ] [ ] [ ] [x] acquisition
[ ] [ ] [ ] [x] legal services
[ ] [ ] [ ] [x] personnel services finance, accounting
[ ] [ ] [ ] [x] infrastructure, archiving and library services
[ ] [ ] [ ] [x] infrastructure, communications (excluding computer communications)
[x] [ ] [ ] [ ] information technology infrastructure, including computer communications, computing, database management, etc.
[ ] [ ] [ ] [x] education
[ ] [ ] [ ] [x] strategic enterprise management
[ ] [ ] [ ] [x] market research
[x] [ ] [ ] [ ] resource management
[x] [ ] [ ] [ ] factory development /development of the production facility)
[ ] [ ] [ ] [x] project management
[ ] [x] [ ] [ ] factory design
[ ] [x] [ ] [ ] factory building and modernization
[ ] [x] [ ] [ ] factory maintenance
[ ] [x] [ ] [ ] infrastructure, energy
[ ] [x] [ ] [ ] infrastructure, buildings and grounds
[ ] [x] [ ] [ ] other (please specify)

11.1.3 What is the highest intended level of genericity on which this architecture is applicable? (check one)

[ ] every type of enterprise
[x] certain type of industry (please specify which one)?

Continuous: steel, paper;Manufacturing: discrete part (electric, electro-mechanical, aerospace) Note: the architecture may be applicable to all types of enterprises.

What is the area of industry in which the architecture has already been applied? *discrete part manufacturing*

11.1.4 Which stage of systems development is addressed by the architecture? (multiple choice question)

[] fully covered	[] partially covered	[] not covered	[] no information	Example

[x] [ ] [ ] [ ] requirements
[x] [ ] [ ] [ ] implementation (configuration and reconfiguration)
[x] [ ] [ ] [ ] design
[x] [ ] [ ] [ ] operation and maintenance

11.1.5 What are the goals which we would like to attain by developing generic CIM architectures? (Which are the main factors of industry's need?)

(Please concentrate on the user requirements of a prospective utilizing industry.)

[] essential factor	[] important factor	[] advantageos but not essential	[] not important	Example

[ ] [ ] [x] [ ] better use of current resources
[x] [ ] [ ] [ ] integration of current technology
[x] [ ] [ ] [ ] cutting excessive costs of having to develop CIM systems individually
[ ] [ ] [x] [ ] improve the quality of the developed individual systems/ manufacturing systems/CIM systems,etc.
[x] [ ] [ ] [ ] enabling the development of a marketplace for compatible CIM products
[x] [ ] [ ] [ ] obtain flexibility in system and adaptation to change in requirements

11.1.6 De facto scope and detail

This question is about how completely the architecture has been defined relative to its intended scope (see Question 2) and its intended detail.

You will have by now a few projects which have already used one part or another of this architecture. We would like to have a list of those activities which were covered by these projects and see what is the level of maturity of the models available (industry mature models, feasibility tested models of individual systems, developed generic models)

Individual case studies – AT&T Network Cable Systems

Development (function, information)
Manufacturing (function, information, resources)
Logistics (function, information, resources)
Manufacturing control – FIAT (FMS)
VW (shop floor control)
Resource management – Bull
IT infrastructrue (communication services, front end services) (+ many other case studies)
Partial models – SW house + University project
(Computer aided enterprise engineering environment/tool) .
Generic models – Fiat & Renault
Enterprise requirements (function, information, resource)
Enterprise system design (function, information).
Enterprise system implementation (function, information)

[The answers given at the time of filling in the questionnaire were not clear with regard to the maturity of the models developed. The Esprit VOICE projects (I and II) were started to validate the CIMOSA modelling framework in various industries]

11.1.7 Omissions

(a) Please check and give details if needed:

[x] the architecture covers its promised intended scope
[ ] the architecture omits some necessary detail (please specify the additional detail the architecture should provide)

(b) If you indicated a need for extension, could you name here any other architecture and model which does cover the missing detail (give specifics) *(Note [Bernus]) In face of the Melbourne Task Force Meeting we should have a mapping between CIMOSA–Purdue–GRAI identifying hitherto unknown missing context/detail (E.g., modelling of the "enterprise building" process).*

11.1.8 Redundancy

Does this architecture require that existing standards must be rewritten (e.g. expressed in a new form without changing the content) in order to fit this architecture?

The architecture is a genuine framework, so constructs and elements which are not part of it have to be modified in order to fill the framework. This work is currently in development for certain levels (design and implementation). However, CIMOSA policy is to adopt existing standards where applicable.

11.1.9 System extensibility

How do you ensure that systems created according to this architecture are open ended?

Existing systems are taken into account in the System Life Cycle by means of a kind of reverse engineering. Thus the modelling framework is filled as if it was a newly designed system. The methodology is both for building new systems and integrating existing ones.

If you judged that the architecture allows modular evolution, is this judgement based on

[x] theoretical considerations
[x] feasibility study or pilot project
[ ] repeated industry experience

11.1.10 Resilience

Does the architecture provide any principles or guidance that will help the user to limit the effect of imposed external changes on the operation of the system?

In terms of application and IT changes (e.g. reconfiguration), and changes in the control flow, the architecture is very adaptable. Similar to the Purdue Architecture there is a separation of function and control. This is the main motivation for distinguishing between "Enterprise Activities" (that describe functionality) and "Business processes" (that describe control by means of "Procedural Rule Sets").

11.1.11 Skills required to implement

Does the architecture adequately treat the skills required by the user to allow for the design and building of a specific system described by it? (give rating):

[x] good[a]

a.The architecture is very good from this respect if the skills required are available or there are tried and natural, pragmatic ways for practitioners to extend their skills in the new direction.

(Note by Bernus) This experience is drawn from electronic discrete part manufacturing. It appears that process industries and discrete part mechanical manufacturing have a different skill base/subculture. This question needs further consideration.

To put it in another way: the electronic industry is more receptive of formal methods/computer science concepts than the rest. It appears that explaining CIMOSA in the Purdue way could make an impact in other industries.

11.1.12 Dependence on new technologies

Does the implementation of this architecture require a given type of new technology for any application within the scope? (A technology is new in this respect if it has not yet been proven under industrial circumstances?)

[ ] no new technology required
[x] yes (please specify what new technology is involved) e.g.
[x] distributed relational databases
[x] distributed object oriented databases
[x] OSI communication
[x] Open Distributed Processing Platform
[x] others (please specify) *Client server technology*

(Note by Bernus) Although other architectures answered 'no', the fact will probably be that they will be materialized on the basis of these new technologies anyway – Therefore the 'yes' answer is not a limitation to the applicability in the manufacturing enterprise.

11.1.13 What is your opinion about the future prospects for this architecture?

[] yes, widespread	[] yes, limited area	[] no	[] undecided

Example

[ ] [ ] [ ] [x] will have theoretical impact
[ ] [x] [ ] [ ] will gain industry acceptance in the present form
[x] [ ] [ ] [ ] will gain industry acceptance in developed (or modified) fo
[x] [ ] [ ] [ ] other (please specify) *will set standards*

11.1.14 What were the reasons for the development of this architecture?

main reason []	cofactor []	not a known reason []	no information []

Example

[ ] [x] [ ] [ ] utilization in a given segment of industry
[ ] [x] [ ] [ ] development of a national standard
[x] [ ] [ ] [ ] development of an international standard
[ ] [x] [ ] [ ] utilization by a given firm
[ ] [x] [ ] [ ] utilization by a group of firms
[ ] [ ] [x] [ ] utilization for teaching purposes
[ ] [ ] [x] [ ] utilization for research purposes (Other, please specify)
[x] [ ] [ ] [ ] (Other please specify): *support integration*
[x] [ ] [ ] [ ] Provide Europe with a competitive technology

11.1.15 What is the driving force behind this architecture?

[] main reason	[] cofactor	[] not a known reason	[] no information

Example

[x] [ ] [ ] [ ] technology pull (there was a perceived need in industry for which the generic architecture intends to be the answer)
[x] [ ] [ ] [ ] technology push (results in the area of architectures methodologies, information technology, communications manufacturing technology created opportunity or incentive for industry to develop this generic CIM architecture)

Please expand your views on the above question if possible.

To provide solutions for the business needs for adapting to changes (internal and external) through system flexibility.

11.1.16 Adequacy

Please rate the adequacy of the architecture with respect to the help which it provides for attaining the essential or important goals which you identified in Question 5.

poor []	average []	good []	excellent []	no info. []

Example

[ ] [ ] [x] [ ] [ ] better use of current resources
[ ] [ ] [ ] [x] [ ] integration of current technology

[ ] [ ] [ ] [x] [ ]	cutting excessive costs of having to develop CIM systems individually
[ ] [ ] [x] [ ] [ ]	improve the quality of the developed individual systems/ manufacturing systems/CIM systems etc.
[ ] [x] [ ] [ ] [ ]	enabling the development of a marketplace for compatible CIM products
[ ] [ ] [x] [ ] [ ]	obtain flexibility in system configuration and adaptation to change in requirements

What is the main strength of this architecture?

- *Providing an overall framework supporting engineering and operation (build and runtime);*
- *Promoting standardization and standards*
- *Enabling enterprise modelling;*
- *Providing enterprise operation flexibility*

To your knowledge which essential or important goals (see Question 5) are specifically not adequately addressed by this architecture? *no answer provided*

11.2 MODELLING

11.2.1 What kinds of models describe this architecture?

Please characterize the models which this architecture requires to be produced? (The table for your answers has been broken into four parts because it had too many columns – see following pages).

Characterize the corresponding modelling languages used to express the models. NB. For the purposes of collecting information we have ordered these tables by the models and not by the modelling languages. Note, that because the same modelling language is often used for more than one model, it is enough to characterize the language only once. However, it is expected that the questions about the language's adequacy for each separate type of model will be answered separately.

Important: the descriptions capturing requirements, designs, etc, are produced in many, very different forms (formal mathematical models, graphs, natural language, tables). We use the word 'model' to refer to all of these forms.

Model[a]	Modelling Language[b]	Degree of formality[c]	Should it be executable?	Is it executable?	Demonstrated feasibility[d]
R[e]	constructs[f]	I → S[g]	partly Y	N → Y	no
D[h]	constructs[i]	I	Y	not yet	no
I[j]	constructs[k]	S[g]	Y	not yet	no

a. Please add rows as necessary. E.g., 'functional requirements model,' 'organizational design model', etc. Note that if models are substantially different on various levels of genericity then you are asked to enter these as separate models.

b. Here 'language' is used in general sense, e.g., ER, Lotos, IDEF0, natural language, Petri net, etc.

c. Please rate: I = Informal, S = Structured but the expressive power of the formal part is not adequate (missing information being added in natural language), F = Formal mathematical model, with formal semantics

d. This is the feasibility of producing an adequate model using this language, P = Practically demonstrated, T = Demonstrated in theory, or only C = Conjectured. Please add further comments in footnotes and on separate pages if necessary.

e. Requirements Specification Model

f. For function/information/resource/organization view respectively IDEF0 + CSP/ERA/–(not yet decided)

g. The CIMOSA Cube with its dimensions

- function/information/resource/organizational views
- requirements/design/implementation
- generic/partial/particular models

h. Design Specification Model

i. Languages on all three levels are similar

j. Implementation Description Model

k. At the moment one language is defined on each level, but the number of languages is still open.

*The Information Integration Infrastructure is modelled in **LOTOS***

Model	Adequacy[a] of modelling language	Is there any related[b] standard modelling language?
R	G → V	see note f previous table
D	G	see note i previous table
I	G (to be further developed)	see note k previous table

a. P = Poor, A = Adequate, G = Good, V = Very good; please comment on problems, limitations in footnote if needed. E.g., overformalized, not formal enough, certain detail not expressible, forces unnecessary decision to be taken too early, etc.

b. If the language used is not a standard and there is a related standard, could you comment, in a footnote, why the selected language is used instead of a standard one?

In all modules, it is believed that the information to be recorded in these models is available and that the languages selected are semantically adequate

Model[a]	Verifiability[b] of model	Could any alternative language be used?	Can completeness be achieved?[c]	Supported by CASE tool?[d]
R	VH → VE	no	N	Y[e]
D	VE	no	N	N
I	VE[f]	no	N[g]	N

a. Please add rows as necessary.
b. Rate: VH = model verifiable by human expert, following a method; VA =automatic or interactive verification is possible; VK = automatic or interactive verification is possible if large knowledge base present; VE = validation possible only by experiments (real or simulation).
c. Y = yes, a complete mathematical model can be developed from which all the necessary design features can be calculated, N = no, – if not, could you tell in a footnote what provision can be made to substitute for the lack of mathematical completeness?
d. Please provide names of CASE tools (or write Y=yes if name is unknown to you), N=no, U=unknown (no information). Note that question 20 will separately ask about the adequacy of the CASE/CAE tools listed.
e. VE is a developed version with simulation for qualification and for release to operation control and manufacturing.
f. CASE Tools for Requirements level are available.
g. Use of simulation to qualify the model.

Due to the size of the models support by CASE tools for verification is needed.

Model[a]	Adequacy of language for people who should use/interpret the model[b]	Adequacy of language for those who should produce the model[c]
R	A	A
D	A	A
I	A	A

a. Please add rows as necessary.
b. Please rate the modelling language P = poor, A = adequate, G = good, V = very good; please comment on problems, limitations in footnote if needed.
c. Please rate the modelling language P = poor, A = adequate, G = good, V = very good; please comment on problems, limitations in footnote if needed.

Due to the size of models CASE tools for the production and interpretation of models are needed. (Note [Bernus]: the original answer by CIMOSA in the table used the word 'compliance' instead of adequacy. Also the CIMOSA answer points out that the CIMOSA languages were designed for end users / business professionals).

11.2.2 How well does the architecture describe the operational characteristics of an enterprise in contrast with describing only its basic design feature?

[x] operational characteristics can be explicitly specified
[x] operational characteristics can be derived from the design features (e.g. through simulation model)

11.2.3 Formalism

Do you think formal methods (with rigorously defined syntax and semantics) are needed to describe and interpret the architecture under discussion and the representation of particular systems by it?

[ ] necessary
[ ] desirable
[x] possible *but limited to design and implementation models*
[ ] not possible

Please comment on what areas you have applied or intend to apply formal methods?...

(Comment [Bernus]): The formal character of the languages should not necessarily be visible to the users. What they see is an ordinary programming language on the Implementation Description level. The formalization takes place at the meta level guaranteeing that the standard excludes misinterpretations and ambiguities.

Could there be a formal canonical view developed together with a modelling language which would subsume all of the ones used? If yes, please specify what should be the characteristics of such a language:

An architecture designed for managing change is hardly conceivable in a canonical form, it must be left open to incorporate user constructs and new business strategies.

Do you think that the scope of the formal methods is broad enough to capture what you want to describe?

[x] no

Can people who need to deal with these models easily relate to these formalisms?

[x] yes

It appears to be that Yes is the answer, but top people are needed to carry out integration projects!

11.2.4 Verification support

How extensively are Computer-aided Engineering (CAE) tools available to verify the prescribed models?

[x] full support for the following models:

Requirements description (function and information view)

Do you think it is necessary to have CAE tools to support the methodology?

Full support means model management, consistency checking of models, and consistency checking between the models produced for different views and levels, and between two successive versions of a model to guarantee smooth online update to the model being executed.

[x] yes, full support is needed

Are there any limitations to achieving the above?

No answer provided

How well can the architecture be used to verify the completeness and adequacy of the overall design for an enterprise or any of its components? (Please comment on merits and problems.)

The goal of CIMOSA is to achieve this.

11.2.5 Economic questions

How well does the associated methodology handle economic questions related to the enterprise described using its methods? (Give comments)

[ ] well covered
[x] some, but not full coverage of issues (what is lacking?)
[ ] not addressed in the modelling
[ ] no information

On the requirements modelling level economic questions are addressed through functional objectives and constraints.

11.3 METHODOLOGY

11.3.1 Does the architecture have a methodology connected to it?

[ ] no
[x] yes (please give the name of the methodology)

CIMOSA Systems Life cycle

If possible, can you specify what the components of the methodology are?

- *Create instances of constructs (G,P,I)*
- *Derivation of models (instantiation)*
- *Generation of views*

How complete is the methodology?

[x] this is a complete methodology, covering all aspects of enterprise engineering activity
[ ] the methodology is a collection of methods/logical techniques (but only separate methodologies are available for the individual modelling tasks)
[ ] the methodology is planned to be a complete one

(Note by Bernus) It appears that the methodological cookbook of CIMOSA has not been written yet, as identified at the Melbourne Task Force Meeting. The Purdue Architecture Mapping on/from CIMOSA or

indirectly the "Purdue Implementation Procedures Manual" for CIMOSA would be a possible methodology.

Does the methodology address the human factor in the enterprise?

[ ] treats the necessary human organization
[ ] skill levels for operating personnel
[ ] training requirements
[ ] treats teams
[ ] treats individual workers

The human factor is described both in the Organizational View in terms of responsibilities and authorities and in the Resource View as skills. Only training is not dealt with explicitly [answer provided without reference to the above choices (Bernus)].

11.3.2 Validation status of the methodology

Was the methodology validated

[ ] in implemented industry projects/systems
[x] in experimental systems
[x] other: (Could you give references?) *Case studies*

References: Contracts – Fiat, Aerospatial, AT&T, Renault, Cap Gemini, NLR,ITGEM,WZL Aachen, etc.

11.3.3 Maturity of the methodology

The methodology[1] has been

[x] successfully taught by others than its developers
[x] successfully applied by other than its developers
[x] repeatedly applied by its developers
[x] shown to work by its developers
[ ] other (e.g. 'under development')

Being taught to University students at TH–Aachen

1. Note [Bernus] as a result of this questionnaire the Task Force had realised how ambiguous the term 'methodology' is in industry. In some cases, like in the answer given here, it means the availability of a modelling framework, in some others it means the availability of detailed work instructions. The distinctions will be dealt with in the matrix mapping of methodologies in this book.

11.3.4 Scope of the methodology

[] extensive coverage	[] some coverage	[] no coverage	[] no information	[] not applicable	Example

Check appropriate fields of the table.

Activities and related methods to	...common to all systems in the scope	...common to a family of systems in the scope (which one?)	...of individual systems
produce requirements...	[x] [ ] [ ] [ ] [ ]	[x] [ ] [ ] [ ] [ ]	[x] [ ] [ ] [ ] [ ]
produce the design specifications...	[x] [ ] [ ] [ ] [ ]	[x] [ ] [ ] [ ] [ ]	[x] [ ] [ ] [ ] [ ]
produce the detailed design...	[x] [ ] [ ] [ ] [ ]	[x] [ ] [ ] [ ] [ ]	[x] [ ] [ ] [ ] [ ]
produce the system...	[ ] [x] [ ] [ ] [ ]	[ ] [x] [ ] [ ] [ ]	[ ] [x] [ ] [ ] [ ]
produce the operational and maintenance methods....	[x] [ ] [ ] [ ] [ ]	[x] [ ] [ ] [ ] [ ]	[x] [ ] [ ] [ ] [ ]

E.g. check 'extensive coverage' box in upper right hand corner if the methodology extensively tackles 'activities and related methods to produce the requirements of individual systems,' to be read from the table as:

Activities and related
methods to

produce the requirements ... ...of individual systems.

Note 1: If the methodology is partially complete in any of these respects pleased add footnote to the box you checked.

Note 2: You can add additional model types below, or on a separate sheet, if you feel that the above categories are not exhaustive:

11.3.5 Presentation of the methodology

[] extensive planned	[] some planned	[] extensive now	[] some now	[] none	[] no info.	[] not applicable

[ ] [ ] [x] [ ] [ ] [ ] [ ] handbook or collection of handbooks
[ ] [ ] [ ] [x] [ ] [ ] [ ] teaching material
[ ] [ ] [ ] [ ] [ ] [x] [ ] organizational procedures
[ ] [ ] [ ] [ ] [ ] [x] [ ] CASE tool support for individuals
[ ] [ ] [x] [ ] [ ] [ ] [ ] project database
[ ] [ ] [x] [ ] [ ] [ ] [ ] CASE tool support for project team

Other form: A PC based Storyboard system is available that helps understand CIMOSA .

11.3.6 Genericity

We want to know how big the gap is between the methodology as presented (see Question 26) and the way it is applied in particular projects.

[x] the methodology describes general procedures only
[x] the methodology is supported by a collection of application examples / cookbook

(Case studies are available) [Note (Bernus): the answers seem a little optimistic. Question: Are there paradigmatic example cases developed to form a cookbook? I propose to say "some individual case studies are available for use."]

11.3.7 Quality of the methodology

Please rate the methodology from the following aspects:

poor []	average []	good []	excellent []	no info. []	N/A []	Example

[ ] [ ] [x] [ ] [ ] [ ] [ ] support of cooperative teamwork
[ ] [ ] [ ] [ ] [x] [ ] [ ] support of work with multiple organizations
[ ] [ ] [ ] [x] [ ] [ ] [ ] flexibility in terms of project size
[ ] [ ] [x] [ ] [ ] [ ] [ ] flexibility in terms of innovativity of project
[ ] [ ] [x] [ ] [ ] [ ] [ ] quality control of project
[ ] [ ] [x] [ ] [ ] [ ] [ ] CAE tools support

12

Evaluation questionnaire for GRAI-GIM

Task Force

12.1 ARCHITECTURE [1] (All answers in italics)

12.1.1 Users of the architecture

What is the intended/targeted audience of this architecture?

Consultants, designers of manufacturing systems in various companies. Any decision maker related to manufacturing systems.

12.1.2 Intended scope of architecture

Please check those activities which are in the intended scope of the architecture (in the short term, i.e. 2-3 years)

[] covered context	[] partially covered context	[] context not covered	[] no information

Example

[ ] [ ] [x] [ ] product research and development
[ ] [ ] [x] [ ] customer services, product maintenance
[ ] [x] [ ] [ ] product management, marketing function, strategic
[ ] [x] [ ] [ ] product distribution
[ ] [ ] [x] [ ] marketing
[ ] [x] [ ] [ ] product design
[x] [ ] [ ] [ ] production planning
[x] [ ] [ ] [ ] manufacturing control
[ ] [x] [ ] [ ] manufacturing

1. Questionnaire filled-in by Peter Bernus based on Questionnaire Version 2; answers provided by Bruno Vallespir and corrected by Guy Doumeingts and Laszlo Nemes at the Toronto Task Force Meeting (May 1992)

[ ] [x] [ ] [ ] material handling, storage and transport
[ ] [x] [ ] [ ] testing and quality control
[ ] [x] [ ] [ ] logistics
[ ] [x] [ ] [ ] acquisition
[ ] [ ] [x] [ ] legal services
[ ] [ ] [x] [ ] personnel services finance, accounting
[ ] [x] [ ] [ ] infrastructure, archiving and library services
[ ] [x] [ ] [ ] infrastructure, communications (excluding computer communications)
[ ] [x] [ ] [ ] information technology infrastructure, including computer communications, computing, database management, etc.
[ ] [x] [ ] [ ] education
[ ] [x] [ ] [ ] strategic enterprise management
[ ] [ ] [x] [ ] market research
[x] [ ] [ ] [ ] resource management
[x] [ ] [ ] [ ] factory development (development of the production facility)
[ ] [x] [ ] [ ] project management
[x] [ ] [ ] [ ] factory design
[ ] [ ] [x] [ ] factory building and modernization
[ ] [x] [ ] [ ] factory maintenance
[ ] [ ] [x] [ ] infrastructure, energy
[ ] [ ] [x] [ ] infrastructure, buildings and grounds
[x] [ ] [ ] [ ] other (please specify): *maintenance*

12.1.3 What is the highest intended level of genericity on which this architecture is applicable? (check one)

[ ] every type of enterprise
[x] certain type of industry (please specify which one)?
(Industrial enterprise)

What is the area of industry in which the architecture has already been applied? *discrete part manufacturing*

12.1.4 Which stage of systems development is addressed by the architecture? (multiple choice question)

[] fully covered	[] partially covered	[] not covered	[] no information	Example

[x] [ ] [ ] [ ] requirements
[x] [ ] [ ] [ ] implementation (configuration and reconfiguration)
[ ] [x] [ ] [ ] design
[ ] [x] [ ] [ ] operation and maintenance

12.1.5 What are the goals which we would like to attain by developing generic CIM architectures? (Which are the main factors of industry's need?)

(Please concentrate on the user requirements of a prospective utilizing industry.)

[] essential factor	[] important factor	[] advanta-geous but not essential	[] not important	Example

[x] [ ] [ ] [ ] better use of current resources
[ ] [x] [ ] [ ] integration of current technology
[x] [ ] [ ] [ ] cutting excessive costs of having to develop CIM systems individually
[x] [ ] [ ] [ ] improve the quality of the developed individual systems/ manufacturing systems/CIM systems, etc.
[ ] [x] [ ] [ ] enabling the development of a marketplace for compatible CIM products
[ ] [ ] [x] [ ] decrease response/product cycle time
[ ] [x] [ ] [ ] contain metrics for exploring economic/technological options
[x] [ ] [ ] [ ] decisional integration
[x] [ ] [ ] [ ] understanding and training
[x] [ ] [ ] [ ] economic aspects
[ ] [x] [ ] [ ] obtain flexibility in system and adaptation to change in requirements
[x] [ ] [ ] [ ] Other (please specify) *develop integration concepts between people*
[x] [ ] [ ] [ ] *promote development of manufacturing system documentation*

12.1.6 De facto scope and detail

This question is about how completely the architecture has been defined relative to its intended scope (see Question 2) and its intended detail.

You will have by now a few projects which have already used one part or another of this architecture. We would like to have a list of those activities which were covered by these projects and see what is the level of maturity of the models available (industry mature models, feasibility tested models of individual systems, developed generic models).

In the view of Guy Doumeingts this question should be reformulated. The following information is as it appeared in the response by Bruno Vallespir (GRAI Lab) in the Version 2 of this questionnaire.

Industry mature models: over a hundred individual production management case studies which produced decisional models of production planning and manufacturing control

Feasibility tested models of individual systems: Production planning (business planning and master production schedule) for electronic assembly of printed wiring boards; Manufacturing control (factory control).

Developed generic models: Information/decision/function aspects of strategic enterprise management

12.1.7 Omissions

(a) Please check and give details if needed):

[ ] the architecture covers its promised intended scope
[ ] the architecture omits some necessary detail (please specify the additional detail the architecture should provide)

GRAI takes into account the economic aspects through its ECOGRAI method. The human aspect is partially/indirectly taken into account in the decisional and organizational models/aspects.

(b) If you indicated a need for extension, could you name here any other architecture and model which does cover the missing detail (give specifics)

It has been suggested that one way to extend GRAI/GIM is to accommodate it in the Purdue Architecture. GRAI Lab (Doumeingts) suggests that this might be one of the possible ways.

12.1.8 Redundancy

Does this architecture require that existing standards must be re-written (e.g. expressed in a new form without changing the content) in order to fit this architecture?

> *We do not know. Perhaps we bring something for standardization in the decisional aspect.*

12.1.9 System extensibility

How do you ensure that systems created according to this architecture are open ended?

> *The architecture was conceived for tackling both new system development and modular evolution or reorganization of old systems into an integrated system.*

If you judged that the architecture allows modular evolution, is this judgement based on

[x] theoretical considerations
[x] feasibility study or pilot project
[x] repeated industry experience

12.1.10 Resilience

Does the architecture provide any principles or guidance that will help the user to limit the effect of imposed external changes on the operation of the system?

> *The effect of changes is taken into account from the technical and from the organizational point of view. GRAI-GIM allows the change of the organization (taking into account possible constraints which exist in the given company).*

12.1.11 Skills required to implement

Does the architecture adequately treat the skills required by the user to allow for the design and building of a specific system described by it? (give rating):

[x] good[a]

a.The architecture is very good from this respect if the skills required are available or there are tried and natural, pragmatic ways for practitioners to extend their skills in the new direction.

12.1.12 Dependence on new technologies

Does the implementation of this architecture require a given type of new technology for any application within the scope? (A technology is new in this respect if it has not yet been proven under industrial circumstances).

[ ] no new technology required
[x] yes (please specify what new technology is involved) e.g.
[x] distributed relational databases
[ ] distributed object oriented databases
[ ] OSI communication
[ ] Open Distributed Processing Platform
[x] others (please specify)

AI techniques (for decision support)

High level fourth generation languages (4 GL)

12.1.13 What is your opinion about the future prospects for this architecture?

[] yes, widespread	[] yes, limited area	[] no	[] undecided

Example

[x] [ ] [ ] [ ] will have theoretical impact
[ ] [x] [ ] [ ] will gain industry acceptance in the present form
[x] [ ] [ ] [ ] will gain industry acceptance in developed (or modified) fo
[x] [ ] [ ] [ ] other (please specify)

Development of an 'Implementation Manual' or 'Users Guide'

12.1.14 What were the reasons for the development of this architecture?

main reason []	cofactor []	not a known reason []	no information []

Example

[ ] [ ] [x] [ ] utilization in a given segment of industry
[ ] [ ] [x] [ ] development of a national standard

[ ] [ ] [x] [ ] development of an international standard
[ ] [ ] [x] [ ] utilization by a given firm
[x] [ ] [ ] [ ] utilization by a group of firms *(1985)*
[ ] [x] [ ] [ ] utilization for teaching purposes
[ ] [ ] [x] [ ] utilization for research purposes *(1977)*
[x] [ ] [ ] [ ] Others: *to solve a problem by a given firm (1980)*

12.1.15 What is the driving force behind this architecture?

[] main reason	[] cofactor	[] not a known reason	[] no information

Example

[x] [ ] [ ] [ ] technology pull (there was a percieved need in industry for which the generic architecture intends to be the answer)
[ ] [x] [ ] [ ] technology push (results in the areas of architectures, methodologies, information technology, communications, manufacturing technology created opportunity or incentive for industry to develop this generic CIM architecture)

Please expand your views on the above quesion if possible.

The GIM architectural approach has developed as an extension to the GRAI method which has been a very successful technique to analyse and specify decisions on multiple levels of the control hierarchy in CIM (production management, planning, control).

12.1.16 Adequacy

Please rate the adequacy of the architecture with respect to the help which it provides for attaining the essential or important goals which you identified in Question 5.

poor []	average []	good []	excellent []	no info. []

Example

[ ] [ ] [x] [ ] [ ] better use of current resources
[ ] [ ] [x] [ ] [ ] integration of current technology
[ ] [ ] [ ] [x] [ ] cutting excessive costs of having to develop CIM systems individually
[ ] [ ] [x] [ ] [ ] improve the quality of the developed individual systems/ manufacturing systems/CIM systems etc.
[x] [ ] [ ] [ ] [ ] enabling the development of a marketplace for compatible CIM products

[ ] [x] [ ] [ ] [ ] decrease response/product cycle time
[ ] [ ] [x] [ ] [ ] contain metrics for exploring economic/technological options
[ ] [ ] [ ] [x] [ ] decisional integration
[ ] [ ] [ ] [x] [ ] understanding and training
[ ] [ ] [x] [ ] [ ] economic aspects
[ ] [ ] [x] [ ] [ ] obtain flexibility in system configuration and adaptation to change in requirements
[x] [ ] [ ] [ ] [ ] other: *develop integration concepts between people*
[x] [ ] [ ] [ ] [ ] *promote development of manufacturing system documentation*

What is the main strength of this architecture?

It suggests an overall, generically applicable structure for designing an integrated Production Management System.

To your knowledge which essential or important goals (see Question 5) are specifically not adequately addressed by this architecture?

Information technology infrastructure.

12.2 MODELLING

12.2.1 What kinds of models describe this architecture?

Please characterize the models which this architecture requires to be produced? (The table for your answers has been broken into four parts because it had too many columns – see following pages).

Characterize the corresponding modelling languages used to express the models. NB. For the purposes of collecting information we have ordered these tables by the models and not by the modelling languages. Note that because the same modelling language is often used for more than one model, it is enough to characterize the language only once. However, it is expected that the questions about the language's adequacy for each separate type of model will be answered separately.

Important: the descriptions capturing requirements, designs, etc, are produced in many, very different forms (formal mathematical models, graphs, natural language, tables). We use the word 'model' to refer to all of these forms.

Model[a]	Modelling Language[b]	Degree of formality[c]	Should it be executable?	Is it executable?	Demonstrated feasibility[d]
0.[e]	GRAI grid	S	Y*		
1.[f]	GRAI grid	S	Y*		
2[g]	GRAICO	S	Y*		
3.[h]	ER	F	Y*		
4.[i]	IDEF0	S	Y*		
5.[j]	IDEF0	S	Y*		
6.[k]	ER	S	Y*		
7.[l]	GRAI Grid	S			
8.[m]	Grai nets	S			
9.[n]	IDEF0[o]	S			

a. Please add rows as necessary. E.g., 'functional requirements model,' 'organizational design model', etc. Note that if models are substantially different on various levels of genericity then you are asked to enter these as separate models.
b. Here 'language' is used in general sense, e.g., ER, Lotos, IDEF0, natural language, Petri net, etc.
c. Please rate: I = Informal, S = Structured but the expressive power of the formal part is not adequate (missing information being added in natural language), F = Formal mathematical model, with formal semantics.
d. This is the feasibility of producing an adequate model using this language, P = Practically demonstrated, T = Demonstrated in theory, or only C = Conjectured. Please add further comments in footnotes and on separate pages if necessary.
e. Global conceptual decisional model
f. Detailed conceptual decisional model
g. Operations model
h. Information model
i. Functional model
j. Conceptual physical model
k. Structural information model
l. Global structural decisional model

m. Detailed structural decisional model
n. Structural physical model
o. IDEF0 with time and synchronization
* Under development

Model	Adequacy[a] of modelling language	Is there any related[b] standard modelling language?
0.[c]	Y	N
1.[c]	Y	N
2.[d]	Y	N
3.[e]	Y	N
4.[f]	Y	N
5.[g]	P	N

a. P = Poor, A = Adequate, G = Good, V = Very good; please comment on problems, limitations in footnote. E.g., over-formalized, not formal enough, certain detail not expressible, forces unnecessary decision to be taken too early, etc.
b. If the language used is not a standard and there is a related standard, could you comment, in a footnote, why the selected language is used instead of a standard one?
c. Apparently there does not exist a standardized language adapted to the decisional model.
d. Petri nets are used but we think that GRAI CO is more complete.
e. We think that the ER model is a pseudo standard therefore we use it.
f. We think that IDEF0 is a pseudo standard therefore we use it.
g. Today there is no precise tool to model the physical system.
2, 3, 4: Although not standardized these languages are widely used in industry.

Model	Verifiability[a] of model	Could any alternative language be used?	Can completeness be achieved?[b]	Supported by CASE tool?[c]
0.	VA	no	N	CAGIM
1.	VH	IDEF0	N	CAGIM
2.	VH	Petri net	N	

3.	VA	NIAM, IDEF1	N	CAGIM and many others
4.	VH	Functional diagram	N	"
5.	VH	Simulation languages	N	"
6.	VH	NIAM, IDEF1	N	"

a. Rate: VH = model verifiable by human expert, following a method; VA =automatic or interactive verification is possible; VK = automatic or interactive verification is possible if large knowledge base present; VE = validation possible only by experiments (real or simulation).
b. Y = yes, a complete mathematical model can be developed from which all necessary design features can be calculated, N = no, if not, could you tell what provision can be made to substitute for the lack of mathematical completeness?
c. Please provide names of CASE tools (or write Y=yes if name is unknown to you), N=no, U=unknown (no information). Note that question 20 will separately ask about the adequacy of the CASE/CAE tools listed.

Model[a]	Adequacy of language for people who should use/interpret the model[b]	Adequacy of language for those who should produce the model[c]
0.	VG	VG
1.	G	G
2.	A	G
3.	A	G
4.	VG	VG
5.	VG	VG

a. Please add rows as necessary.
b. Please rate the modelling language P = poor, A = adequate, G = good, V = very good; please comment on problems, limitations in footnote if needed.
c. Please rate the modelling language P = poor, A = adequate, G = good, V = very good; please comment on problems, limitations in footnote if needed.

12.2.2 How well does the architecture describe the operational characteristics of an enterprise in contrast with describing only its basic design feature?

[x] operational characteristics can be derived from the design features (e.g. through simulation model)

12.2.3 Formalism

Do you think formal methods (with rigorously defined syntax and semantics) are needed to describe and interpret the architecture under discussion and the representation of particular systems by it?

[ ] necessary
[ ] desirable
[ ] possible
[x] not possible *at the moment*

Please comment on what areas you have applied or intend to apply formal methods?

As apparent from answers to Question 17, the information model is based on ER which can be considered as a formal model.

Could there be a formal canonical view developed together with a modelling language which would subsume all of the ones used? If yes, please specify what should be the characteristics of such a language:

Manufacturing System modelling needs two levels, descriptive and formal. Starting from reality we develop a descriptive model first (to start with a formal model seems to be too difficult). In our approach using our modelling tools, we develop a model of the reality. The objective is to simplify but to keep the main features. From this descriptive model we elaborate a formal model in order to improve the design. We have used this approach for simulation i.e. first we developed our IDEF0 model, then 'Activities' model (CSL-CAPS). We are developing our research in this direction.

Do you think that the scope of the formal methods is broad enough to capture what you want to describe?

No answer provided

Can people who need to deal with these models easily relate to these formalisms?

No answer provided

12.2.4 Verification support

How extensively are Computer-aided Engineering (CAE) tools available to verify the prescribed models?

[x] some support for the following models: *All, limited: GRAICO*

Do you think it is necessary to have CAE tools to support the methodology?

[x] yes, full support is needed

Are there any limitations to achieving the above?

No answer provided.

How well can the architecture be used to verify the completeness and adequacy of the overall design for an enterprise or any of its components? (Please comment on merits and problems.)

There is no conclusive evidence yet. We know that the architecture is good at the conceptual and structural levels.

12.2.5 Economic questions

How well does the associated methodology handle economic questions related to the enterprise described using its methods? (Give comments)

[x] well covered
[ ] some, but not full coverage of issues (what is lacking?)
[ ] not addressed in the modelling
[ ] no information

Comments: *Economic Tableau ('Tableaux de Board'). Economic evaluation tools currently being developed.*

12.3 METHODOLOGY

12.3.1 Does the architecture have a methodology connected to it?

[ ] no
[x] yes (please give the name of the methodology) *GIM*

If possible, can you specify what are the components of the methodology?

1. Overall decisional modelling

2. Detailed decisional modelling

3. Physical view design
4. Information systems design
5. User-oriented design
6. Technically oriented design

How complete is the methodology?

- [ ] this is a complete methodology, covering all aspects of enterprise engineering activity
- [x] the methodology is a collection of methods/logical techniques (but only separate methodologies are available for the individual modelling tasks)
- [ ] the methodology is planned to be a complete one

Does the methodology address the human factor in the enterprise?

- [x] treats the necessary human organization
- [ ] skill levels for operating personnel
- [x] training requirements
- [x] treats teams *partially*
- [ ] treats individual workers

12.3.2 Validation status of the methodology

Was the methodology validated

- [x] in implemented industry projects/systems
- [ ] in experimental systems
- [ ] other: (Could you give references?)

No answer provided

12.3.3 Maturity of the methodology

The methodology has been

- [x] successfully taught by others than its developers
- [x] successfully applied by other than its developers
- [x] repeatedly applied by its developers
- [x] shown to work by its developers
- [ ] other (e.g. 'under development')

12.3.4 Scope of the methodology

[] extensive coverage	[] some coverage	[] no coverage	[] no information	[] not applicable	Example

Check appropriate fields of the table.

Activities and related methods to ↙	...common to all systems in the scope	...common to a family of systems in the scope (which one?)	...of individual systems
produce requirements...	[x] [ ] [ ] [ ] [ ]	[x] [ ] [ ] [ ] [ ]	[x] [ ] [ ] [ ] [ ]
produce the design specifications...	[x] [ ] [ ] [ ] [ ]	[x] [ ] [ ] [ ] [ ]	[x] [ ] [ ] [ ] [ ]
produce the detailed design...	[ ] [ ] [ ] [ ] [x]	[ ] [ ] [ ] [ ] [x]	[ ] [ ] [ ] [ ] [x]
produce the system...	[ ] [x] [ ] [ ] [ ]	[ ] [x] [ ] [ ] [ ]	[ ] [x] [ ] [ ] [ ]
produce the operational and maintenance methods....	[x] [ ] [ ] [ ] [ ]	[x] [ ] [ ] [ ] [ ]	[x] [ ] [ ] [ ] [ ]

Example: check 'extensive coverage' box in upper right hand corner if the methodology extensively tackles 'activities and related methods to produce the requirements of individual systems,' to be read from the table as:

Activities and related
methods to ↘
produce the requirements ... ...of individual systems.

Note 1: If the methodology is partially complete in any of these respects pleased add footnote to the box you checked.

Note 2: You can add additional model types below, or on a separate sheet, if you feel that the above categories are not exhaustive:

12.3.5 Presentation of the methodology

[] extensive planned	[] some planned	[] extensive now	[] some now	[] none	[] no info.	[] not applicable

[ ] [ ] [x] [ ] [ ] [ ] [ ] handbook or collection of handbooks
[ ] [ ] [ ] [x] [ ] [ ] [ ] teaching material
[ ] [ ] [x] [ ] [ ] [ ] [ ] organizational procedures
[ ] [ ] [x] [ ] [ ] [ ] [ ] CASE tool support for individuals
[ ] [ ] [x] [ ] [ ] [ ] [ ] project database
[ ] [ ] [ ] [x] [ ] [ ] [ ] CASE tool support for project team
[ ] [ ] [ ] [x] [ ] [ ] [ ] any other form? *metaformalism to describe the* modelling languages

12.3.6 Genericity

We want to know how big the gap is between the methodology as presented (see Question 26) and the way it is applied in particular projects.

[x] the methodology describes general procedures only
[ ] the methodology is supported by a collection of application examples / cookbook (*In the future*)

12.3.7 Quality of the methodology

Please rate the methodology from the following aspects:

poor []	average []	good []	excellent []	no info. []	N/A []	Example

[ ] [ ] [ ] [x] [ ] [ ] [ ] support of cooperative teamwork
[ ] [ ] [ ] [x] [ ] [ ] [ ] support of work with multiple organizations
[ ] [ ] [ ] [x] [ ] [ ] [ ] flexibility in terms of project size
[ ] [ ] [x] [ ] [ ] [ ] [ ] flexibility in terms of innovativity of project
[ ] [ ] [ ] [ ] [ ] [ ] [x] quality control of project *(not yet)*
[ ] [ ] [x] [ ] [ ] [ ] [ ] CAE tools support *(not yet)*

13

Evaluation questionnaire for PERA

Task Force

13.1 ARCHITECTURE [1] (All answers in italics)

13.1.1 Users of the architecture

What is the intended/targeted audience of this architecture?

No answer provided

13.1.2 Intended scope of architecture

Please check those activities which are in the intended scope of the architecture (in the short term, i.e. 2–3 years)

[] covered context	[] partially covered context	[] context not covered	[] no information

Example

[ ] [ ] [ ] [ ] product research and development
[ ] [ ] [ ] [ ] customer services, product maintenance
[ ] [ ] [ ] [ ] product management, marketing function, strategic
[ ] [x] [ ] [ ] product distribution
[ ] [ ] [ ] [ ] marketing
[ ] [x] [ ] [ ] product design
[ ] [x] [ ] [ ] production planning
[ ] [x] [ ] [ ] manufacturing control
[ ] [x] [ ] [ ] manufacturing
[ ] [x] [ ] [ ] material handling, storage and transport

1. Questionnaire filled in by Peter Bernus after the Melbourne Task Force Meeting, May 1992; answers are based on T. J. Williams' answers to Questionnaire Version 2 and on discussions at the Melbourne Task Force Meeting.

[ ] [x] [ ] [ ] testing and quality control
[ ] [x] [ ] [ ] logistics
[ ] [x] [ ] [ ] acquisition
[ ] [ ] [ ] [ ] legal services
[ ] [ ] [ ] [ ] personnel services finance, accounting
[ ] [ ] [ ] [ ] infrastructure, archiving and library services
[ ] [x] [ ] [ ] infrastructure, communications (excluding computer communications)
[ ] [x] [ ] [ ] information technology infrastructure, including computer communications, computing, database management, etc.
[ ] [x] [ ] [ ] education
[ ] [x] [ ] [ ] strategic enterprise management
[ ] [ ] [ ] [ ] market research
[ ] [ ] [ ] [ ] resource management
[ ] [ ] [ ] [ ] factory development /development of the production facility)
[ ] [ ] [ ] [ ] project management
[ ] [ ] [ ] [ ] factory design
[ ] [ ] [ ] [ ] factory building and modernization
[ ] [ ] [ ] [ ] factory maintenance
[ ] [ ] [ ] [ ] infrastructure, energy
[ ] [ ] [ ] [ ] infrastructure, buildings and grounds
[ ] [ ] [ ] [ ] other (please specify)

Note: *where there is no 'x', answer was not provided*

13.1.3 What is the highest intended level of genericity on which this architecture is applicable? (check one)

[x] every type of enterprise
[ ] certain type of industry (please specify which one)?

What is the area of industry in which the architecture has already been applied? *The architecture is too new for it to have been applied yet*

13.1.4 Which stage of systems development is addressed by the architecture? (multiple choice question)

[] fully covered	[] partially covered	[] not covered	[] no information	Example

[x] [ ] [ ] [ ] requirements
[x] [ ] [ ] [ ] implementation (configuration and reconfiguration)

[x] [ ] [ ] [ ] design
[x] [ ] [ ] [ ] operation and maintenance

13.1.5 What are the goals which we would like to attain by developing generic CIM architectures? (Which are the main factors of industry's need?)

(Please concentrate on the user requirements of a prospective utilizing industry.)

[] essential factor	[] important factor	[] advantageous but not essential	[] not important	Example

[x] [ ] [ ] [ ] better use of current resources
[x] [ ] [ ] [ ] integration of current technology
[ ] [ ] [ ] [ ] cutting excessive costs of having to develop CIM systems individually
[x] [ ] [ ] [ ] improve the quality of the developed individual systems/ manufacturing systems/CIM systems,etc.
[ ] [ ] [x] [ ] enabling the development of a marketplace for compatible CIM products
[ ] [ ] [ ] [ ] obtain flexibility in system and adaptation to change in requirements

13.1.6 De facto scope and detail

This question is about how completely the architecture has been defined relative to its intended scope (see Question 2) and its intended detail.

No sufficient information available at this stage

13.1.7 Omissions

(a) Please check and give details if needed):

[x] the architecture covers its promised intended scope
[ ] the architecture omits some necessary detail (please specify the additional detail the architecture should provide)

(b) If you indicated a need for extension, could you name here any other architecture and model which does cover the missing detail (give specifics).

13.1.8 Redundancy

Does this architecture require that existing standards must be rewritten (e.g. expressed in a new form without changing the content) in order to fit this architecture?

No

13.1.9 System extensibility

How do you ensure that systems created according to this architecture are open ended?

No answer provided.

If you judged that the architecture allows modular evolution, is this judgement based on

[ ] theoretical considerations
[ ] feasibility study or pilot project
[x] repeated industry experience *[of the authors (Bernus)]*

13.1.10 Resilience

Does the architecture provide any principles or guidance that will help the user to limit the effect of imposed external changes on the operation of the system?

The master plan, which describes the integrated architecture, provides a means for analyzing and locating the effects of change. Also resilience of individual systems in the face of technological changes is a feature of these systems which can be designed.

The division in the architecture, of control and material flow is one of the principles that separate the actual technology from the rest of the system. Thus as far as functional requirements are met technology changes have a localized effect. The same is true of the decision of how to mix human and automated functions.

13.1.11 Skills required to implement

Does the architecture adequately treat the skills required by the user to allow for the design and building of a specific system described by it? (give rating):

[x] very good[a]

a.The architecture is very good from this respect if the skills required are available or there are tried and natural, pragmatic ways for practitioners to extend their skills in the new direction.

The presentation of the architecture is apparently very well matching the view of industrial systems as perceived by engineers and managers. This has been verified in the process industry.

13.1.12 Dependence on new technologies

Does the implementation of this architecture require a given type of new technology for any application within the scope? (A technology is new in this respect if it has not yet been proven under industrial circumstances)?

[x] no new technology required

13.1.13 What is your opinion about the future prospects for this architecture?

[] yes, widespread	[] yes, limited area	[] no	[] undecided

Example

[ ] [ ] [ ] [x] will have theoretical impact
[x] [ ] [ ] [ ] will gain industry acceptance in the present form
[x] [ ] [ ] [ ] will gain industry acceptance in developed (or modified) form
As evidenced by the Purdue CIM consortium

13.1.14 What were the reasons for the development of this architecture?

main reason []	cofactor []	not a known reason []	no information []

Example

[ ] [ ] [x] [ ] utilization in a given segment of industry
[ ] [ ] [x] [ ] development of a national standard
[ ] [ ] [x] [ ] development of an international standard
[ ] [ ] [x] [ ] utilization by a given firm
[x] [ ] [ ] [ ] utilization by a group of firms
[x] [ ] [ ] [ ] utilization for teaching purposes
[x] [ ] [ ] [ ] utilization for research purposes (Other, please specify)

13.1.15 What is the driving force behind this architecture?

[] main reason	[] cofactor	[] not a known reason	[] no information

Example

[x] [ ] [ ] [ ] technology pull (there was a perceived need in industry for which the generic architecture intends to be the answer)

[ ] [x] [ ] [ ] technology push (results in the area of architectures methodologies, information technology, communications manufacturing technology created opportunity or incentive for for industry to develop this generic CIM architecture)

Please expand your views on the above quesion if possible.

The success of the Purdue CIM Model for industrial control in the area of continuous industries has created the initiative to extend this concept to

- *any type of industry*
- *all activities of the enterprise, not only process control*

This cofactor made it possible to fill in the genuine need for an enterprise integration architecture and accompanying methodology.

13.1.16 Adequacy

Please rate the adequacy of the architecture with respect to the help which it provides for attaining the essential or important goals which you identified in Question 5.

There is not sufficient information in this respect (yet)

What is the main strength of this architecture?

Its graphical and easy to grasp presentation

The philosophy of how it divides concerns is meeting the way of thinking of industry practitioners

To your knowledge which essential or important goals (see Question 5) are specifically not adequately addressed by this architecture? *Unknown*

13.2 MODELLING

13.2.1 What kinds of models describe this architecture?

Please characterize the models which this architecture requires to be produced? (The table for your answers has been broken into four parts because it had too many columns – see following pages).

Characterize the corresponding modelling languages used to express the models. NB. For the purposes of collecting information we have ordered these tables by the models and not by the modelling languages. Note, that because the same modelling language is often used for more than one model, it is enough to characterize the language only once. However, it is expected that the questions about the language's adequacy for each separate type of model will be answered separately.

The descriptions capturing requirements, designs, etc, are produced in many, very different, forms (formal mathematical models, graphs, natural language, tables) We use the word 'model' to refer to all of these forms.

Overall plans are incomplete at this point.

There is no commitment in the Purdue Architecture as to how (or whether) to develop formal models; at the moment it is up to the team creating the master plan what way of description suits best the given individual system's architecture.

Overall note: it appears that the Purdue architecture develops to great detail the architecture proper, while other approaches, like CIMOSA and GRAI, provide modelling framework and modelling methods to fill in the elements of the Purdue Architecture.

Model[a]	Modelling Language[b]	Degree of formality[c]	Should it be executable?	Is it executable?	Demonstrated feasibility[d]
0.[e]					
1.[f]					
2[g]	Overall plans are incomplete at this point.				
3.[h]					
4.[i]					

5.[j]	
6.[k]	
7.[l]	Overall plans are incomplete at this point.
8.[m]	
9.[n]	
10.[o]	
11.[p]	

a. Please add rows as necessary. E.g., 'functional requirements model,' 'organizational design model', etc. Note that if models are substantially different on various levels of genericity then you are asked to enter these as separate models.
b. Here add rows as necessary. E.g., 'functional requirements model,' 'organizational design model', etc. Note that if models are substantially different on various levels of genericity then you are asked to enter these as separate models.
c. Please rate: I = Informal, S = Structured but the expressive power of the formal part is not adequate (missing information being added in natural language), F = Formal mathematical model, with formal semantics.
d. This is the feasibility of producing an adequate model using this language, P = Practically demonstrated, T = Demonstrated in theory, or only C = Conjectured Please add further comments in footnotes and on separate pages if necessary.
e. Requirements for control
f. Management vision
g. Requirements for customer services (CS)
h. Task description (control)
i. Task description
j. Task and functional module description (control)
k. Task and functinal module description (CS)
l. Information flow between tasks (control)
m. Task interconnection (material and information) customer service
n. Information system architecture
o. Human and organizational architecture
p. Customer response architecture

Note: (e)–(m) belong to the definition of the system architecture while (n)–(p) are repeated for each of specification, detailed design, manifestation and operations. Levels of description of the systems' architecture for items ((e)–(m) include management information, production scheduling anf planning, intraarea coordination, supervisory control, direct digital control, dedicated controllers and process.

Model	Adequacy of modelling language	Is there any related standard modelling language?
0–11[a]	*No sufficient information yet*	

a. *see models e–p*

Model	Verifiabilityof model	Could any alternative language be used?	Can completeness be achieved?	Supported by CASE tool?
0–11[a]	*No sufficient information yet*			

a. *see models e–p*

Model	Adequacy of language for people who should use/interpret the model	Adequacy of language for those who should produce the model
0–11[a]	*No sufficient information yet*	

a. *see models e–p*

13.2.2 How well does the architecture describe the operational characteristics of an enterprise in contrast with describing only its basic design feature?

[x] operational characteristics can be explicitly specified
[ ] operational characteristics can be derived from the design features (e.g. through simulation model)

The links are explicitly described in the Master Plan, however, automatic derivation (especially because of no committment to use formal models) is not possible.

13.2.3 Formalism

Do you think formal methods (with rigorously defined syntax and semantics) are needed to describe and interpret the architecture under discussion and the representation of particular systems by it?

[x] not necessary in general

Please comment on what areas you have applied or intend to apply formal methods?...

This view may change later, however at present there is no full evidence apart from specific subsystems (E.g., critical software modules need to be formally specified and verified).

Could there be a formal canonical view developed together with a modelling language which would subsume all of the ones used? If yes, please specify what should be the characteristics of such a language:

Unknown at present, in the lack of sufficient experimental evidence.

Do you think that the scope of the formal methods is broad enough to capture what you want to describe?

No answer provided

Can people who need to deal with these models easily relate to these formalisms?

No answer provided

13.2.4 Verification support

How extensively are Computer Aided Engineering (CAE) tools available to verify the prescribed models? *No answer provided*

Do you think it is necessary to have CAE tools to support the methodolog

No answer provided

Are there any limitations to achieving the above? *No answer provided*

How well can the architecture be used to verify the completeness and adequacy of the overall design for an enterprise or any of its components? (Please comment on merits and problems.)

The verification of the architectures based on the generic Purdue Architecture is a human design process. The Implementation Procedures Manual, which is the extensive cookbook style description of how to write a master plan can be followed to ensure complete coverage of the subject.

13.2.5 Economic questions

How well does the associated methodology handle economic questions related to the enterprise described using its methods? (Give comments)

[x] well covered

13.3 METHODOLOGY

13.3.1 Does the architecture have a methodology connected to it?

[x] yes (please give the name of the methodology) *Purdue Methodology*

If possible, can you specify what are the components of the methodology?

The 'Purdue Architecture' is the base framework for the 'Implementation Procedures Manual' (IPM). The IPM corresponds in structure to the Master Plan describing in detail how a Master Plan can be created.

How complete is the methodology?

[x] this is a complete methodology, covering all aspects of enterprise engineering activity
[ ] the methodology is a collection of methods/logical techniques (but only separate methodologies are available for the individual modelling tasks)
[ ] the methodology is planned to be a complete one

Does the methodology address the human factor in the enterprise?

[x] treats the necessary human organization
[x] skill levels for operating personnel
[x] training requirements
[x] treats teams
[x] treats individual workers

13.3.2 Validation status of the methodology

Was the methodology validated

[x] in implemented industry projects/systems
[x] other: *Case studies. Accepted by potential users*

Could you give references? *No answer provided*

13.3.3 Maturity of the methodology

The methodology has been

[x] successfully taught by others than its developers
[x] successfully applied by other than its developers
[x] repeatedly applied by its developers
[x] shown to work by its developers

13.3.4 Scope of the methodology

[] extensive coverage	[] some coverage	[] no coverage	[] no information	[] not applicable	Example

Check appropriate fields of the table.

Activities and related methods to ↙	...common to all systems in the scope	...common to a family of systems in the scope (which one?)[a]	...of individual systems
produce requirements...	[ ] [x] [ ] [ ] [ ]	[ ] [ ] [ ] [ ] [ ]	[ ] [x] [ ] [ ] [ ]
produce the design specifications...	[ ] [x] [ ] [ ] [ ]	[ ] [ ] [ ] [ ] [ ]	[ ] [x] [ ] [ ] [ ]
produce the detailed design...	[ ] [x] [ ] [ ] [ ]	[ ] [ ] [ ] [ ] [ ]	[ ] [x] [ ] [ ] [ ]
produce the system...	[ ] [x] [ ] [ ] [ ]	[ ] [ ] [ ] [ ] [ ]	[ ] [x] [ ] [ ] [ ]
produce the operational and maintenance methods....	[ ] [x] [ ] [ ] [ ]	[ ] [ ] [ ] [ ] [ ]	[ ] [x] [ ] [ ] [ ]

a. *The methodology does not tackle 'partial models' (a term borrowed from CIMOSA). However the architecture may accommodate the concept of partial model/description. Partial and particular models flow from the generic Purdue Models.*

Example: check 'extensive coverage' box in upper right hand corner if the methodology extensively tackles 'activities and related methods to produce the requirements of individual systems,' to be read from the table as:

Activities and related
methods to ↪
produce the requirements ... ...of individual systems.

13.3.5 Presentation of the methodology

[] extensive planned	[] some planned	[] extensive now	[] some now	[] none	[] no info.	[] not applicable

[ ] [ ] [x] [ ] [ ] [ ] [ ] handbook or collection of handbooks
[ ] [ ] [x] [ ] [ ] [ ] [ ] teaching material
[ ] [ ] [x] [ ] [ ] [ ] [ ] organizational procedures
[ ] [ ] [ ] [ ] [x] [ ] [ ] CASE tool support for individuals
[ ] [ ] [ ] [x] [ ] [ ] [ ] project database
[ ] [ ] [ ] [ ] [x] [ ] [ ] CASE tool support for project team

13.3.6 Genericity

We want to know how big the gap is between the methodology as presented (see Question 26) and the way it is applied in particular projects.

[x] the methodology describes general procedures only
[ ] the methodology is supported by a collection of application examples / cookbook

13.3.7 Quality of the methodology

Please rate the methodology from the following aspects:

poor []	average []	good []	excellent []	no info. []	N/A []	Example

Since this methodology is new, there is no implementation experience as (as of 1992)

14

Analysis of questionnaire results

P. Bernus and L. Nemes

The conclusions below are drawn from a review of the answers given to each of the three Questionnaires by the Purdue, CIMOSA and GRAI-GIM teams.

14.1 ARCHITECTURE

14.1.1 Users of the architecture

The architectures are targeted at diverse audience, such as

1. Consultants, Engineering Firms
2. Designers of the enterprise (technical level)
3. Decision makers involved in the design and development of the enterprise such as:
   (a) Projects to integrate enterprise activities (carried out as a project within the enterprise
   (b) Commissioned projects (e.g., for the delivery of a subsystem to the enterprise such as a manufacturing system or any other customer service.)

Note: generic architectures are to be handled by many decision makers who might be at various levels in the decision making hierarchy.

14.1.2 Intended scope of the architecture

We have asked what area of enterprise activity is generally involved in generic architecture proposals. The answer to this question is twofold:

- What areas of activity are intended to be covered
- What areas have already been covered and to what extent have they been covered.

In other words we wanted to determine the potential of the architecture versus the actual state of its use as of today.

The first question is, in a way, more important to answer than the second but watching closely

- What is the time frame in which the missing details can in fact be delivered.
- Are the methods already provided with the architecture satisfactory enough so that we can expect that the not yet covered areas will indeed be tackled in the same quality as the already worked out details have been.

As a result of these ponderables the following conclusions can be drawn.

- The following areas of enterprise activity are common to all the architectures:

  Product distribution, production planning, manufacturing control, manufacturing, material handling, storage and transport, testing and quality control, logistics, acquisition, information technology infrastructure, (including computer, communications) computing, database management systems, etc.
- Those areas which are common to some but not all:

  Product design, infrastructure, communications (excluding computer communications), education, strategic enterprise management, resource management, factory development (i.e. development, of the production facility), factory design, factory maintenance.
- Listed only by Purdue:

  Legal services, market research.
- Listed only by CIMOSA:

  Product research and development, factory building and modernization, infrastructure, energy, infrastructure, buildings and grounds.
- Listed only by GRAI-GIM:

  Product management, marketing function, project management.
- These items not covered by any of the architectures (or the issue has not yet been addressed):

  Customer services, product maintenance, marketing.

Enterprise integration architectures have grown out of manufacturing integration architectures and methodologies, thus their initial scope is the same as for manufacturing integration. The actual expertise and worked out details are in this area.

However, as we shall see later these enterprise approaches look at more aspects even of the manufacturing related activities than the earlier manufacturing integration projects did.

14.1.3 What is the highest intended level of genericity on which this architecture is applicable?

Every type of industrial enterprise can be tackled using these generic architectures, but to date the actual work on details has concentrated on those industries which have the characteristics of discrete parts manufacturing or of continuous process manufacturing (electronic equipment, paper, chemical, food processing, mechanical and electro mechanical part and machine manufacturing, car industry).

It appears, however, that the service industries can also be tackled by these architectures (e.g., engineering firms, software firms, etc.).

14.1.4 Which stage of systems development is addressed by the architecture?

By intention all stages of systems development are fully covered, thus requirements, design, implementation (configuration and reconfiguration), operation and maintenance. 'Being covered by the architecture' should mean that the architecture must describe

- generic models for the enterprise (at each level of the development process), and
- generic models of the processes involved in producing the above.

It is a major source of confusion that each of the three architectures uses a different terminology for the above concepts.

The Purdue and GRAI-GIM way of presenting the architecture is based on the engineer's view, (1) what project specific major deliverables are there? and (2) how to produce them?

The CIMOSA way of presenting the architectures is according to an abstract categorization of models.

The modelling methods of these architectures allow the entire life history of the integrated enterprise to be covered.

Generic models for this are available in the form of the CIMOSA life cycle model, or the Purdue Architecture diagram.

14.1.5 What are the goals which we would like to attain by developing generic CIM architectures? (What are the main factors in industry's need?)

It was felt by the groups answering these questions that the important and essential factors leading to the development of generic architectures are:[1]

** Better use of current resources

*** Integration of current technology

** Cutting excessive costs of having to develop enterprise integration systems individually

** Improve the quality of the developed individual enterprises/ manufacturing systems/CIM systems, etc. (goal orientedness)

** Enabling the development of a marketplace for compatible products for integration

* Contain metrics for exploring economic/technological options

* Integration of the decision making process

* Obtain flexibility in system configuration and adaptation to change in requirements

* To develop for people involved the concept of the integrated nature of the business. This includes understanding, awareness, and training.

* Development of a documented form of the enterprise processes. This is an essential factor for assessing and controlling the quality of the process and thereby the quality of the product.

14.1.6 De facto scope and detail

14.1.6.1 Case studies

A number of individual case studies have been prepared for both GRAI-GIM and CIMOSA. The Purdue architecture has also been applied to individual cases in industry although it is very new, however, the Purdue Reference Model which concentrates on the manufacturing and manufacturing control areas was tested on numerous industry cases. In fact the CIMOSA and GRAI-GIM case studies have also addressed different subparts of the scope.

1. The number of *s show how many have felt that the factor was essential.

14.1.6.2 Generic architectural models

The generic architectures in the abstract have almost all the enterprise activities in scope. Detailed generic models which can be used as cookbook examples are at the moment not available except for a few cases, such as:

(a) Available

- The Purdue reference model for manufacturing control
- Production Planning and Production Control Models produced by the GRAI method
- Strategic Enterprise Management Models produced by the GRAI method

(b) Under Development

- Computer Aided Enterprise Engineering Model/Environment (a software house/university project of CIMOSA)
- Fiat/Renault generic enterprise requirements/system design/implementation models.

The CIMOSA Integrating Infrastructure is a generic model for the information integrating infrastructure and is intended to be developed using ISO and other standards (such as OSI (Open Systems Interconnection), ODP (Open Distributed Processing) or RDA (Remote Database Access), and user interface standards etc.) Note that various similar projects exist in theU.S., Canada and in the Asia-Pacific region and it is unlikely that any one will become completely dominant.

The importance of these standards is that the existence of a concrete example for a generic standard is essential for industry acceptance.

14.1.7 Omissions

- The importance of equal treatment of economic and human aspects has only recently been acknowledged by the more technically oriented methodologies. Today, the intentions of these architectures is to be as complete as possible.
- The economic aspect of enterprise integration has improved in the last 8 years, e.g., the GRAI methodology includes the ECOGRAI method.
- The Purdue architecture explicitly treats the human aspect.

14.1.8 Interaction of the modelling framework selected and existing standards expressed in different modelling terms

CIMOSA has adopted a uniform modelling framework and the effect of the above problem is currently being investigated.

The Purdue architecture does not use a fixed modelling framework, so the problem does not arise.

The GRAI-GIM architecture is based on widely accepted existing modelling methods. Where not (in its detailed conceptual decisional model and the operational model) there are modelling methods which are quasi standard (IDEF0 and Petri nets). Where GRAI adds a new method (GRAI-grid) there are no standards available and it is hoped that one of GRAI's inputs to CIMOSA will be the development of CIMOSA's decisional/organizational aspects.

14.1.9 Systems extensibility

In each of the architectures the necessity of extending existing systems has been taken into account.

There are subtle differences, e.g. the Purdue architecture proposes to remodel the existing part of the system only after the new system has been functionally defined to a certain level.

The CIMOSA life-cycle model proposes to fill the modelling framework, using reverse engineering methods, for the existing system.

Proof that the transition from old to new systems can be done using the considered architectures is based on a mixture of theoretical consideration, feasibility studies and repeated industry experience (Purdue, GRAI).

It is therefore important that the life cycle models of CIMOSA and the Purdue Architecture description explicitly give methods to describe this migration strategy. However, a detailed migration strategy to deal with legacy systems (application programs and databases) is yet to be developed for each architecture, especially for migration of mission critical systems.

14.1.10 Resilience of systems using the generic architectures

Perhaps the most important message of the evaluation is that changeability (which underlies resilience) is a designed property of the enterprise or any of its subsystems.

A generally accepted structural division, used explicitly in both Purdue and CIMOSA, is the division of the control functions from the controlled functions. This gives the system a general quality of localizing the effects of change. E.g., the same elementary technological processes may be entirely rearranged using a different control system and still the elementary processes may remain the same. This is important given the different phases and speed of technological shifts in the computing/control technology and manufacturing technology.

14.1.11 Skills required to implement

The architectures appear to be very different from this aspect. This involves the language, style, and presupposed knowledge on behalf of the practitioner.

The Purdue architecture is easily understood by industrial engineers while the CIMOSA and GRAI-GIM architectures need some information technology background for ready understanding.

14.1.12 Development potential with respect to skills to implement

CIMOSA intends to finalize the modelling framework in a way that hides from the practitioner the underlying theoretical complexity.

Purdue, when its practical engineering approach is mapped with the GIM or CIMOSA life cycle models, will become theoretically under-pinned and it is expected that the three architectures can be merged.

The electronic industry is more receptive of formal methods using abstract models than is the process industry. It is therefore expected that the application of CIMOSA and GRAI to process industry case studies and the application of the Purdue Architecture to electronic industry case studies will bring out the fundamental correspondences as well as to help identify mutual contribution between them.

14.1.13 Dependence on new technologies

It was the opinion of the majority that for enterprise integration architectures to be practicable, there will be an extensive need for a number of new information technologies.

Distributed Databases (relational and heterogeneous), Open Systems Interconnection, Open Distributed Processing, AI Techniques, Fourth Generation Languages were all mentioned as important conditions for practical implementations.

Distributed information technology is a currently changing R and D area and too early standardization of higher levels is not likely to be long lived.

There was no one technology mentioned, however, which seems to lock the user of these architectures into the use of that technology alone.

One strategic question arises concerning the Information Integration Infrastructure (IIS) of CIMOSA. It should be assessed to what extent the specifications of IIS will accommodate (be resilient to) the obvious spread of distributed AI techniques in the area of information integration.

14.1.14 What is your opinion about the future prospects for this architecture?

Answers to this question by the developers of the three architectures are hard to evaluate. CIMOSA claims the setting of standards, presumably referring to the rigorous modelling and specification methods. Purdue claims industry acceptance in its present form, backed by the Purdue consortium. GRAI claims a theoretical impact (by introducing the decisional aspect and its modelling methods) and influencing the other two architectures to develop in this direction. Also the ECOGRAI method may evolve to become an influential part of generic architectures and methodologies.

14.1.15 What has been the main reason for the development of this architecture?

The Purdue and GRAI Architectures were initially based on academic developments with industry support and utilization by a group of firms.

CIMOSA is a product of industrial research, backed by academia, but the main reason has been for the development of a standard and for utilization by a group of firms.

Today each of the architectures is backed by a group of firms.

14.1.16 What is the driving force behind this architecture?

The major reason for the development of generic architectures seems to be technology pull, i.e. on the basis of industrial needs rather than on earlier developed technology being used in a new area of application.

This has created a situation in which rapid, partial development is easily made. It is then very hard to withhold results until they are really finished and under-pinned with a clear theoretical base. The reason for the creation of this Task Force is a symptom of the above situation.

Thus, enterprises utilizing the generic architectures need to understand that they are technologies which are still developing.

The decision of an enterprise to apply these architectures should be based on strategic decisions because the expected benefits can only be assured if the maturity level of the enterprise is raised to a sufficient extent.

14.1.17 Adequacy

The strong points of the CIMOSA architecture are the following:
- It is developing an overall framework supporting engineering and application (i.e. designing and building the enterprise) called the Information Integration Infrastructure
- It promotes standardization with concrete standardizable modelling languages
- It enables enterprise modelling in the above framework.

The strong points of the Purdue Architecture are
- Its easy to grasp presentation
- The philosophy on how it divides the views of the enterprise matches the way people think of (and organize) enterprises. This philosophy is eminently suitable for use by industrial practitioners.

The strength of the GRAI/GIM Architecture
- GRAI suggests a universally applicable approach to structure production management.
- The GRAI-GIM architecture allows an explicit modelling and design of the decisional structure of manufacturing systems, while most of the existing approaches are more information processing oriented.

14.1.18 Further conclusions regarding the architectures

In addition to the conclusions drawn from the individual questions, here are a few more comments:
1. The most important contribution of the Purdue Architecture is its detailed and practical proposal as to how to go about integrating enterprises. Some aspects (like manufacturing control) are dealt with in great detail.
2. The model and description categories of the Purdue architecture are defined in a finer detail than in the other two architectures (see the answers to Question 17). Furthermore there are detailed methods for producing all these descriptions provided one has selected a specific modelling method (or English) to capture these descriptions.

3. The important CIMOSA contribution is to define a generic integration infrastructure (IIS) although it is not known to the authors at the moment how available the LOTOS specification is for IIS.
4. Although IIS depends on yet to be fixed international standards it was announced that prototypical implementations are already underway. The nature of the IIS's dependence is therefore not more than the dependence on any other type of operating system with communication and multiprocessing facilities. In fact it is only anticipated that the IIS will have an ODP implementation.
5. The other important CIMOSA contribution is the definition of its modelling framework. With some amendments (to be treated in conjunctions with the Matrix Method [see Chapter 17]) it might be usable with Purdue and other methodologies.
6. A possibility not yet explored of the CIMOSA modelling framework is its application to the integration process itself. This would enable the use of a great number of software technology results and make it possible to apply, in conjunction with CIMOSA or the other methodologies, software process models/standards.
7. The GRAI-GIM architecture is structurally closer to the Purdue architecture than CIMOSA in the sense that it changes the number of views/aspects after the detailed design models (This change is called 'user oriented' vs. 'technical' division in GRAI-GIM terminology).
8. The GRAI-GIM modelling framework is very close to CIMOSA, with the exception that GRAI-GIM uses basically standard methods while CIMOSA uses a single tailor made one. GRAI-GIM architectural models are therefore easier to support with a diversity of CASE tools.
9. We believe that the number of viewpoints (aspects) as defined in CIMOSA is bound to expand in the future and that the development of matching (CIMOSA compliant) modelling languages instead of existing ones may prove to be a bottleneck. Note that modelling languages take 10 –15 years to mature and there is no shortcut known.
10. The CIMOSA approach gains however, an easier coordination of the different aspects as more formal cross model checking is made possible.
11. A major lack of every architecture studied is that they put forward life cycle models which involve only those processes which directly contribute to the transformation of various forms and levels of enterprise descriptions. The life cycle models need to be expanded to include models of the 'transition from computer supported business to the information integrated enterprise.' Many processes necessary to this transition would then get represented and coordinated with the rest even though

they might not directly contribute to the transformation of architecture documents.

14.1.19 Vision of the future, for further study

We foresee the following tasks ahead:

1. Map the architectures onto one another in order to identify in detail the correspondence between the processes and models described by each.
2. Prepare case studies to apply CIMOSA, GRAI-GIM and Purdue to the same (kind of) project in order to compare the results. Also a case study to use the Purdue Implementation Procedures Manual but with CIMOSA and GRAI models could validate or disprove the conjecture that these combinations could significantly advance the state-of-the-art.
3. Produce a generic model for the 'transition from the computer supported business to the information integrated enterprise' as a supplement to a pure life cycle model.
4. Evaluate the use of emerging techniques as possible components of the IIS (like distributed databases, heterogeneous databases, multimedia communication, distributed AI, etc.)
5. Develop a combined architecture, possibly consisting of the generic enterprise integration process (a la Purdue) and the descriptions produced using the combined CIMOSA and GRAI model categories.
6. The most detailed correspondence between the candidate architectures could be unveiled by developing the respective meta models for these architectures.

15

Short form evaluation questionnaire for integration reference architectures

J. Vlietstra

This short questionnaire was developed to provide a quick overview of the questions tackled below in the detailed questionnaire.

1. Coverage

a	technical aspects
b	business
c	human relations
d	human skills

2. Modelling support

a	functions
b	information
c	resources
d	organization
e	social aspects
f	economics
g	legalities
h	finance & administration

3. Support description methods or languages for each of the items of question 2?
4. Support derivation methods to derive functional operations and functional entities from the descriptions of question 2?

5. Provide an engineering environment?
6. Provide an operation environment?
7. Provide simulation tools?
8. Comply with international standards?
9. Does the architecture have the sympathy of vendors?

15.1 ANSWERS TO SHORT FORM QUESTIONNAIRE FOR CIMOSA

Table 15.1 Answers to short form questionnaire for CIMOSA

1. Coverage		
a	technical aspects	yes
b	business	yes
c	human relations	no
d	human skills	no
2. Modelling support		
a	functions	yes
b	information	yes
c	resources	yes
d	organization	yes
e	social aspects	via functions
f	economics	via functions
g	legalities	via functions
h	finance and administration	via functions
3. Descriptive methods support		(first four only a–d)
4. Derivation methods support		(see question 2)
5. Engineering environment		yes
6. Operations environment		yes
7. Simulation tools		use established ones
8. Complies with international standards		this is intended; no international standard on CIM is available at present.
9. Has been proven		yes
10. Has substantial vendor support		yes, substantial

15.2 ANSWERS TO SHORT FORM QUESTIONNAIRE FOR GRAI-GIM

Table 15.2 Answers to short form questionnaire for GRAI-GIM

1. Coverage		
a	technical aspects	yes
b	business	yes
c	human relations	yes, partially
d	human skills	yes, partially
2. Modelling support		
a	functions	yes
b	information	yes
c	resources	yes
d	organization	yes
e	social aspects	no
f	economics	no
g	legalities	no
h	finance and administration	no
3. Descriptive methods support		(first four only a–d)
4. Derivation methods support		(see question 2)
5. Engineering environment		yes, CAGIM tools
6. Operations environment		not in the sense of CIMOSA
7. Simulation tools		use established ones
8. Complies with international standards		this is intended; no international standard on CIM is available at present.
9. Has been proven		yes, through industrial applications
10. Has substantial vendor support		yes, several companies support

15.3 ANSWERS TO SHORT FORM QUESTIONNAIRE FOR PURDUE

Table 15.3 Answers to short form questionnaire for Purdue

1. Coverage		
a	technical aspects	yes
b	business	yes
c	human relations	yes
d	human skills	yes
2. Modelling support		
a	functions	yes
b	information	yes
c	resources	yes
d	organization	yes
e	social aspects	yes
f	economics	via project evolution methods
g	legalities	no
h	finance and administration	via functions
3. Descriptive methods support		(all listed in item 2)
4. Derivation methods support		(all listed in item 2)
5. Engineering environment		yes
6. Operations environment		no
7. Simulation tools		use established ones
8. Complies with international standards		as desired by implementor when these standards become available
9. Has been proven		yes
10. Has substantial vendor support		yes

16

Techniques to map the architectures directly against one another

T.J. Williams, D. Zoetekouw, J.P. Shewchuk, D. Chen and H. Li

16.1 SOME INITIAL OBSERVATIONS

Before beginning a discussion of the mapping techniques actually used by the Task Force in its work, it is useful to discuss the techniques of mapping themselves, particularly their negative aspects. It should thus be noted that:

- Any form of direct comparison may show one architecture in a better light than the others.
- Nobody wants to look bad, particularly the developers of these architectures.
- Mappings, matrices, etc., as used in this section are all direct comparison techniques.
- We originally proposed the mapping techniques at the first meeting in Bordeaux to determine the overlaps of the architecture – not which was best in any way – and to determine whether any or all of them would give a complete coverage of the field of enterprise development – to furthermore determine the holes in the coverage for future developments.
- We should be treating the current candidates as a collective resource, not as competitors in a contest for architectural supremacy.
- We got into this situation because we were using each other's basic diagram as a set of templates for its relative mapping. This was easiest but has led to the current competition. The format should be neutral.

What is really needed is the following:

- A matrix of needs (fulfilled and unfulfilled) should be developed. Those needs which one or more of the current candidates (or others) has fulfilled should then be indicated.

- The unfulfilled needs remaining then become the basis for future research and development.
- Continuing work should then propose ways in which these needs might be accomplished.
- In other words, we should be collectively designing the overall needed architectures and related techniques.

At the same time, we should note that it is impossible to show everything in one relatively small set of diagrams. Thus many of our collective diagrams and techniques may together be helpful to users. They should make the choice, not us.

16.2 A COMPARISON OF VIEWS POINTING OUT THE SEMANTIC PROBLEM

The next few drawings, Figures 16.1–6, present the relationships of several terms used by each of the architectures to indicate the various analysis, design, implementation, construction and operation functions as they occur during the life history of any major project. The first two are between the terms used by CIMOSA and Purdue, the next two between GRAI-GIM and Purdue and the last pair between CIMOSA and GRAI-GIM. These figures

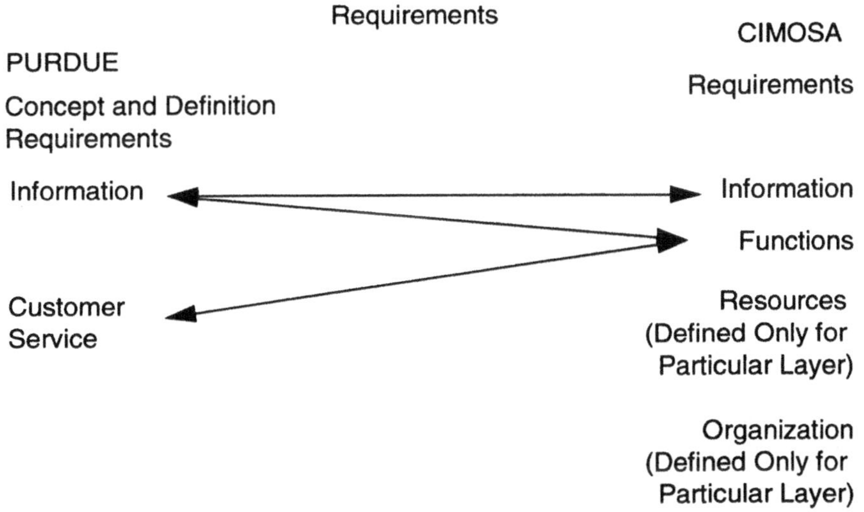

Figure 16.1 Comparison of CIMOSA and PURDUE semantics for the requirements task.

quickly show the considerable differences in meaning given to the same words or phrases by the different groups or conversely how the different groups call the same operations by different terms. These diagrams also readily show by their multitude of interconnecting lines, the major difficulties which will be involved in mapping one architecture onto any other.

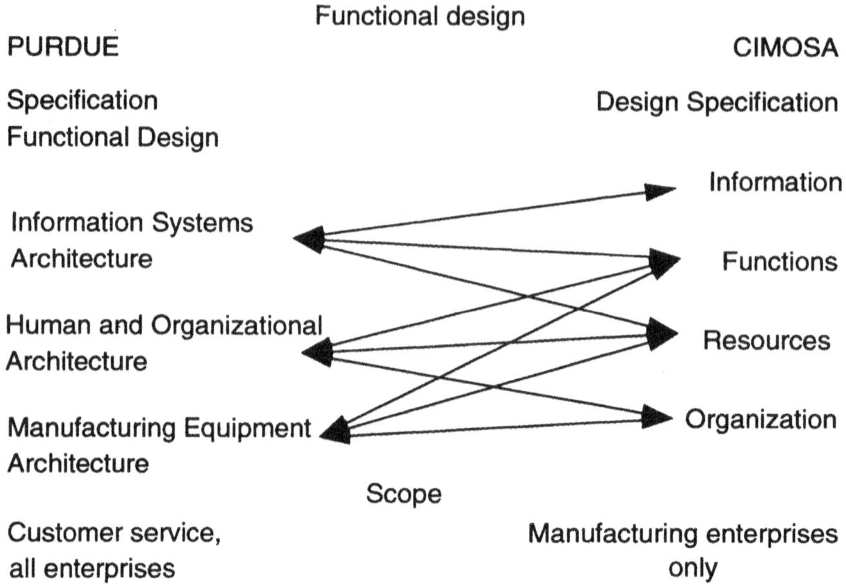

Figure 16.2 Comparison of CIMOSA and PURDUE semantics for the functional design task.

16.3 MAPPING AGAINST OTHER ARCHITECTURES

During 1991–92 there was considerable discussion concerning how a mapping of the characteristics of any one of the architecture against the others could be carried out. Many ideas were expressed but all were discarded for various reasons, mostly because they would have been too complex to convey useful information. The chosen methods are discussed below.

The result of these discussions is a method based upon a corepresentation of the 'life history' diagrams of each architecture. Where any particular stage of the life history was not shown by one of the architectures this would be

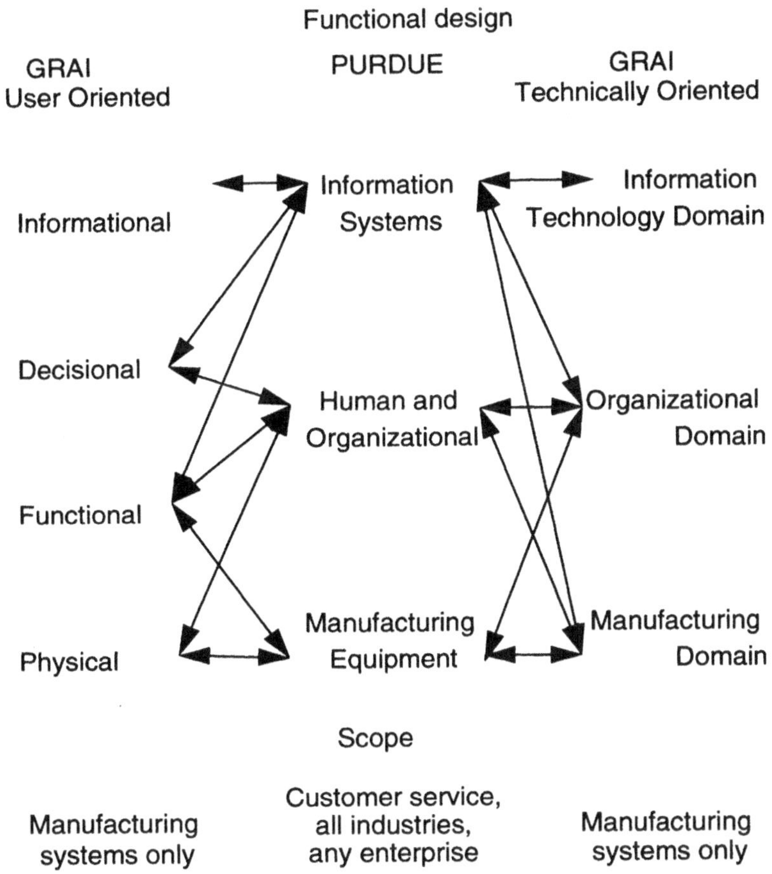

Figure 16.3 Comparison of GRAI–GIM and PURDUE semantics for the functional design task.

added to the diagram for that architecture. This is illustrated by describing the CIMOSA representation.

As sketched in Fig. 16.7, this begins by taking a side view from the right of any one vertical Layer of the CIMOSA Modelling Framework. The result is shown in Fig. 16.8. To the above representation three additional horizontal blocks are added. These are entitled, (1), the Identification and Concept Block, (2) Build, and (3) Operation. They then give a complete Life History representation in the diagram.

The example Layer of the Modelling Framework plus the added Identification and Concept Block becomes a larger four level block entitled

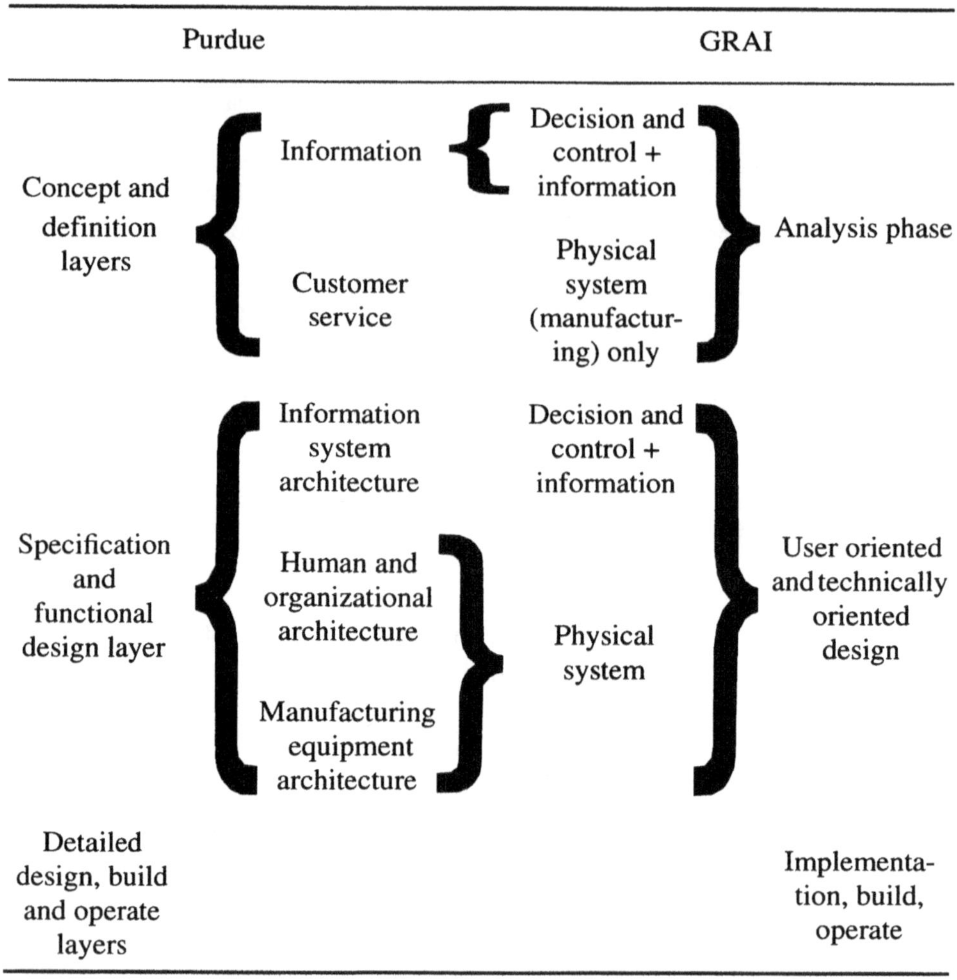

Figure 16.4 Comparison of the sematics of Purdue and GRAI-GIM.

Analysis and Development. The designators of the subdivisions of this block are given in Table 16.1, entitled List of Abbreviations. The Purdue diagram (Fig. 16.11) is the Structural Model of Figure 8.1 of Chapter 8. The GRAI-GIM diagram (Fig. 16.12) is the GIM Structured Approach of Figure 7.19 of Chapter 7.

The individual architectures are then mapped against each of the above diagrams. Fig. 16.9 shows the meaning of each area of the CIMOSA diagram as represented by the CIMOSA Architecture itself. The diagram concludes that CIMOSA provides no mechanism for the determination and modelling

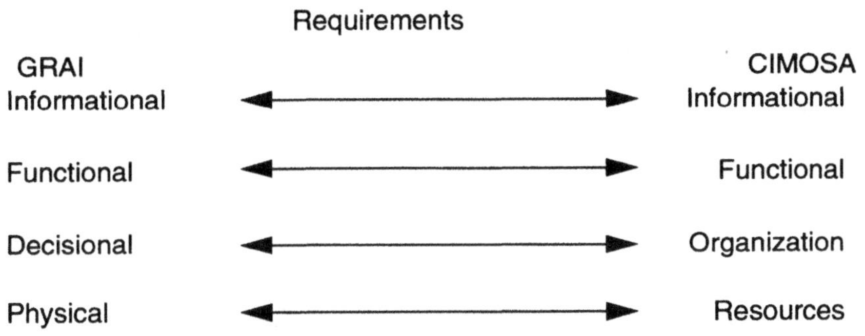

Figure 16.5 Comparison of the semantics for GRAI-GIM and CIMOSA for the requirements task.

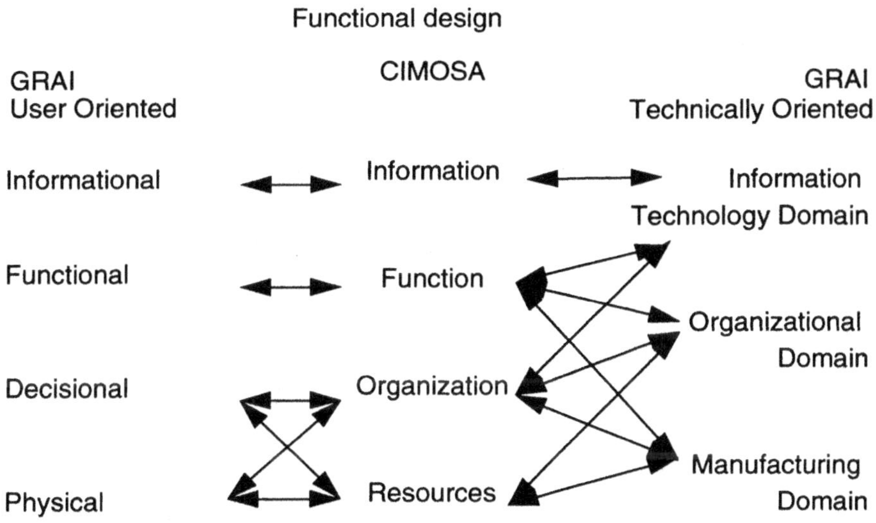

Figure 16.6 Comparison of the semantics of CIMOSA, GRAI-GIM and PURDUE for the functional design task.

of what Purdue calls Identification and Concept or what GRAI-GIM calls User Requirements and Initialization, ie, management's reasons for getting into an integration project in the first place, their goals and objectives and the proposed scope of the overall project. In CIMOSA these are assumed to be available prior to the initialization of the project and it is considered that the mechanism of their acquisition is not part of the CIMOSA scope.

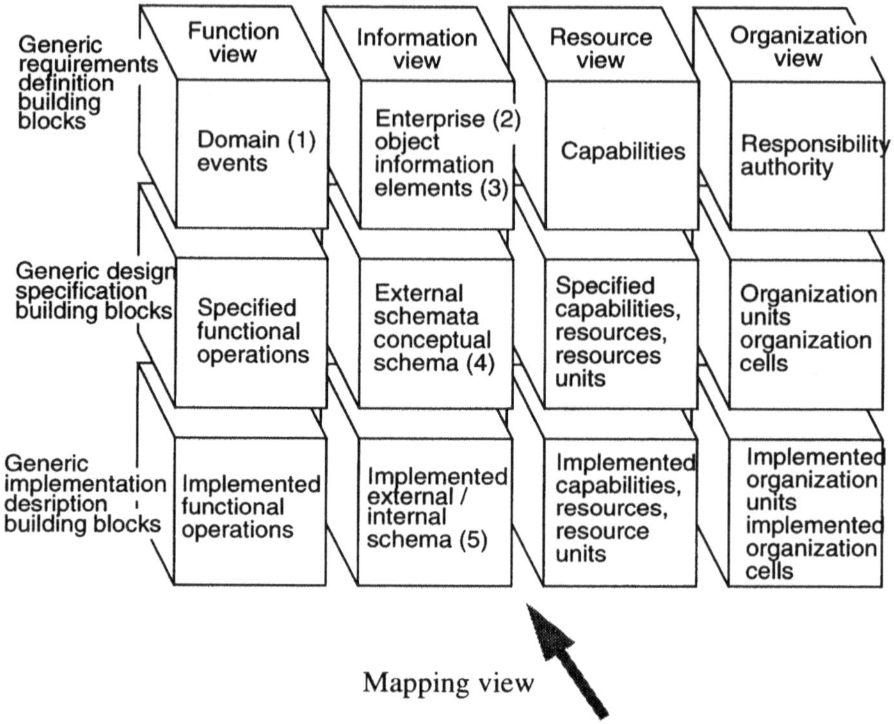

Figure 16.7 Overview of CIMOSA generic building blocks
(1) domain processes, enterprise activities; (2) object views, object relationships; (3) integrity rules; (4) integrity constraints, database transactions; (5) logical / physical database schemata.

Likewise while the Build process is discussed it is not modelled. The Operate block is defined by the Application, Human and Manufacturing Functional Entities respectively.

The footnotes define the extent of the scope of CIMOSA in these areas. The information presented in the top and left hand side of Fig. 16.9 defines the CIMOSA tools and methodology which are applicable at the designated locations on the diagram.

Fig. 16.10 presents the mapping of PERA onto the CIMOSA Basic Skeleton Diagram. Several things are immediately evident.

- The Identification and Concept Block is utilized as noted above.
- The separation of enterprise tasks by Purdue into those related to the Information System versus the Customer Service System is indicated on

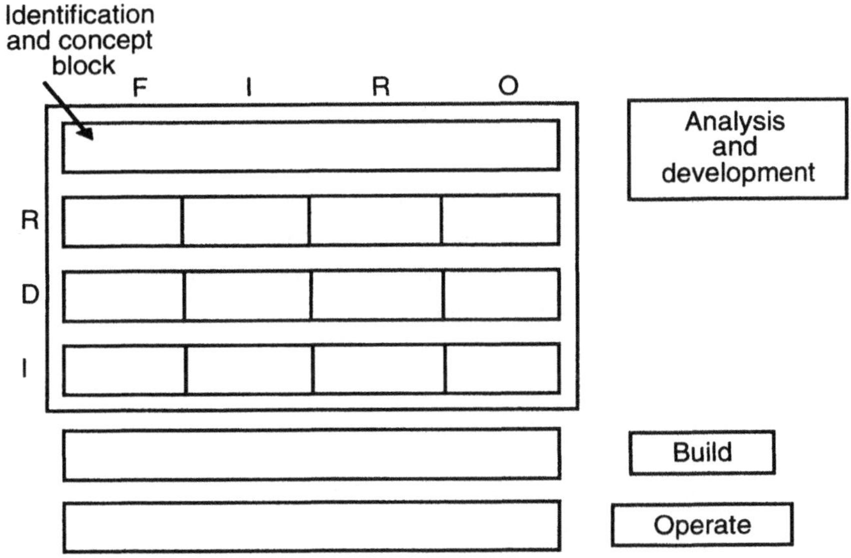

Figure 16.8 Development of the CIMOSA basic skeleton diagram.

the diagram by splitting each block in the Modelling Framework horizontally to show this extra capability.

- Similarly, the separate consideration of and emphasis on Human Tasks in both the Information and Customer Service areas is indicated on the Resource (R) blocks of the Design (D) and Implementation (I) Layers, Note that the distinction between Information and Customer Service for the human is maintained only for the Design Level (Specification in Purdue).
- At Implementation (Detailed Design in Purdue) all human considerations are consolidated together as the Human and Organizational Architecture.
- Purdue gives considerable coverage to the BUILD process (Construction, Checkout and Commissioning) of the integrated enterprise, called Manifestation in Fig. 16.11.
- Likewise, Purdue also extensively covers the optimal operation of the integrated enterprise (OPERATE) including the concepts of Continuous Improvement and Total Quality Management (TQM).

Fig. 16.13 presents the mapping of GRAI-GIM onto the CIMOSA Basic Skeleton Diagram. As noted below this mapping was described by the developers of GRAI-GIM themselves. Again several basic points can be made concerning this mapping.

Table 16.1 List of abbreviations used on architecture mappings

Symbol	*Explanation*
IIS	Integrated Infrastructure
IEEE	Integrated Enterprise Engineering Environment
IEOE	Integrated Enterprise Operation Environment
F,I,R,O	Function, information, resource and organizational views
AFE	Application Functional Entity
HFE	Human Functional Entity
MFE	Machine Functional Entity
R,D,I	Requirements, design and implementation levels
OPE	Operational level
CI/HI	Computer/Human implementation information
HCS	Human implementation customer service
MCS	Machine implementation customer service
INF	Information
CS	Customer service
● PURDUE Mapped Purdue capabilities on CIMOSA and GRAI-GIM	
#CIMOSA Mapped CIMOSA capabilities on Purdue and GRAI-GIM	
* GRAI-GIM Mapped GRAI-GIM capabilities on Purdue and CIMOSA	

- GRAI-GIM does not claim a Build or Operate capability. It is apparently confined to the Requirement, Design and Implementation phases of CIMOSA or the corresponding Functional Requirements, Functional Design or Specification, and Detailed Design Phases of Purdue.
- A more detailed discussion of this mapping is presented in the next section of this Chapter.

Figures 16.14 and 16.15 present the mappings of CIMOSA and GRAI-GIM onto the structure of the Purdue Enterprise Reference Architecture. In view of Figures 16.9 and 16.10 the explanation is probably straightforward. However, the structure of Fig. 16.15 requires the explanation of Fig. 16.21 and 16.22 for the symbology used. Perusal of the further explanation in the GRAI paper ie: Section 16.4 of this Chapter, will probably also be helpful. This material will not be duplicated here in the interests of space. Fig. 16.16 shows a mapping of CIMOSA and Purdue on the GRAI-GIM Structured Approach Diagram. Again, in view of the proceeding material this diagram should be self explanatory.

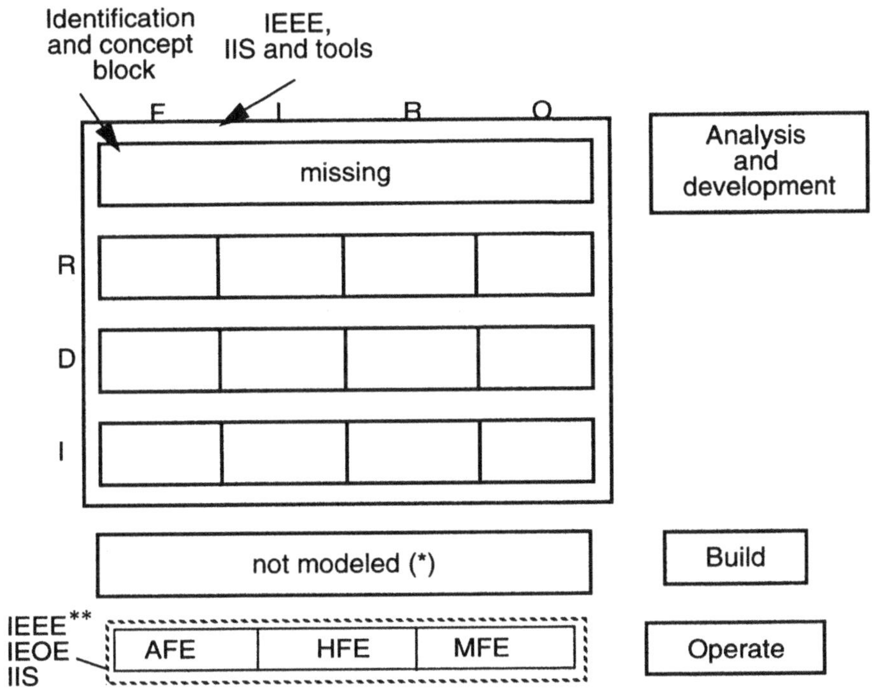

Figure 16.9 CIMOSA basic skeleton diagram CIMOSA on CIMOSA (IEEE:Integrated Enterprise Engineering Environment; IIS: Integrating Infrastructure; IEOE: Integrated Enterprise Operation Environment; AFE: Application ~, HFE: Human ~, MFE: Machine Functional Entity) * Process and tools are available for the modelling; ** Systems change addressed only for applications software change and development.

16.4 MAPPING OF GRAI-GIM AGAINST[1] CIMOSA AND PURDUE

There was no opportunity during any of the scheduled meetings of the Task Force when the developers of the GRAI-GIM architecture and methodology could directly voice their considerations concerning the various one-on-one mappings. This was because they were unable to attend any of the same

1. The introduction and definitions of this section on 'Mapping of GRAI-GIM Against CIMOSA and Purdue' were prepared by G. Doumeingts, B. Vallespir, D. Chen, M. Zanetten, and F. Marcotti of GRAI/LAP, University of Bordeaux. The part of Mapping of GRAI-GIM onto the Purdue Enterprise Reference Architecture was prepared by J. Shewchuk, of Purdue, who spent the Fall Semester 1992 at the GRAI/LAP Laboratory. This work is dated December 5, 1992.

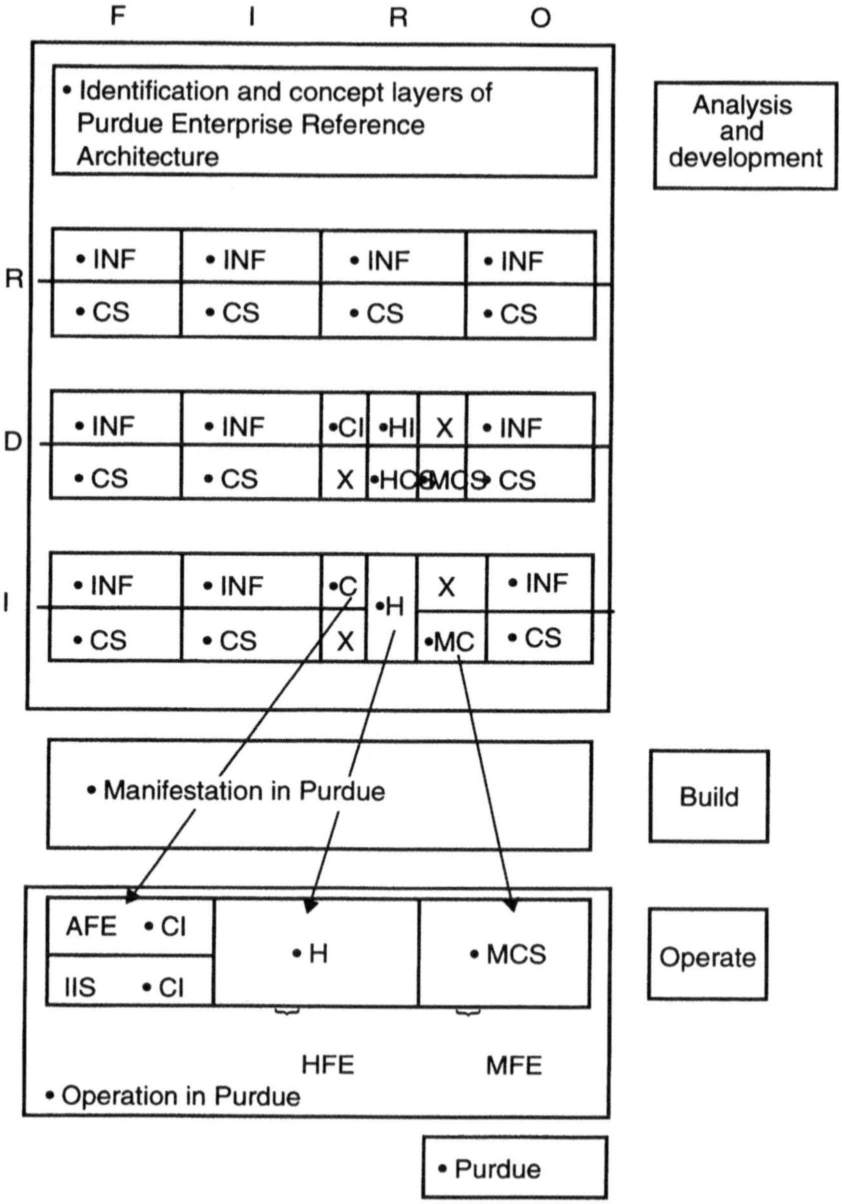

Figure 16-10 Mapping of the capabilities of PERA on the CIMOSA architecture.

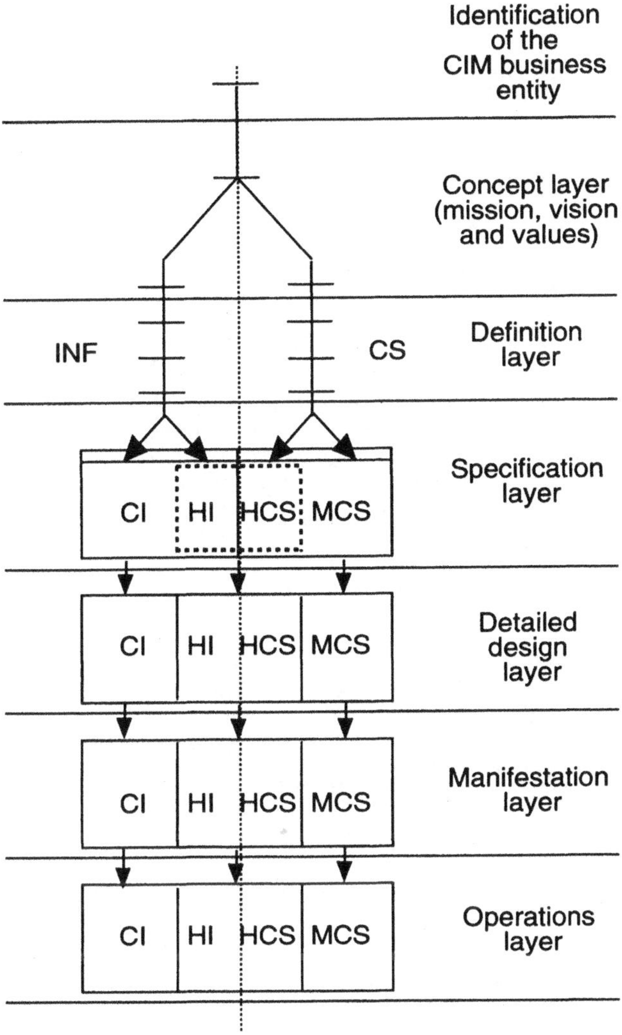

Figure 16-11 Elemental structural diagram: Purdue Enterprise Reference Architecture.

meetings with the developers of the other architectures after these mappings techniques had been proposed. As a result the only way such a comparison could be obtained was for the GRAI group to develop a separate paper concerning their opinions on this subject. This paper is presented herewith as a separate section of this chapter. The paper will be presented in its entirety

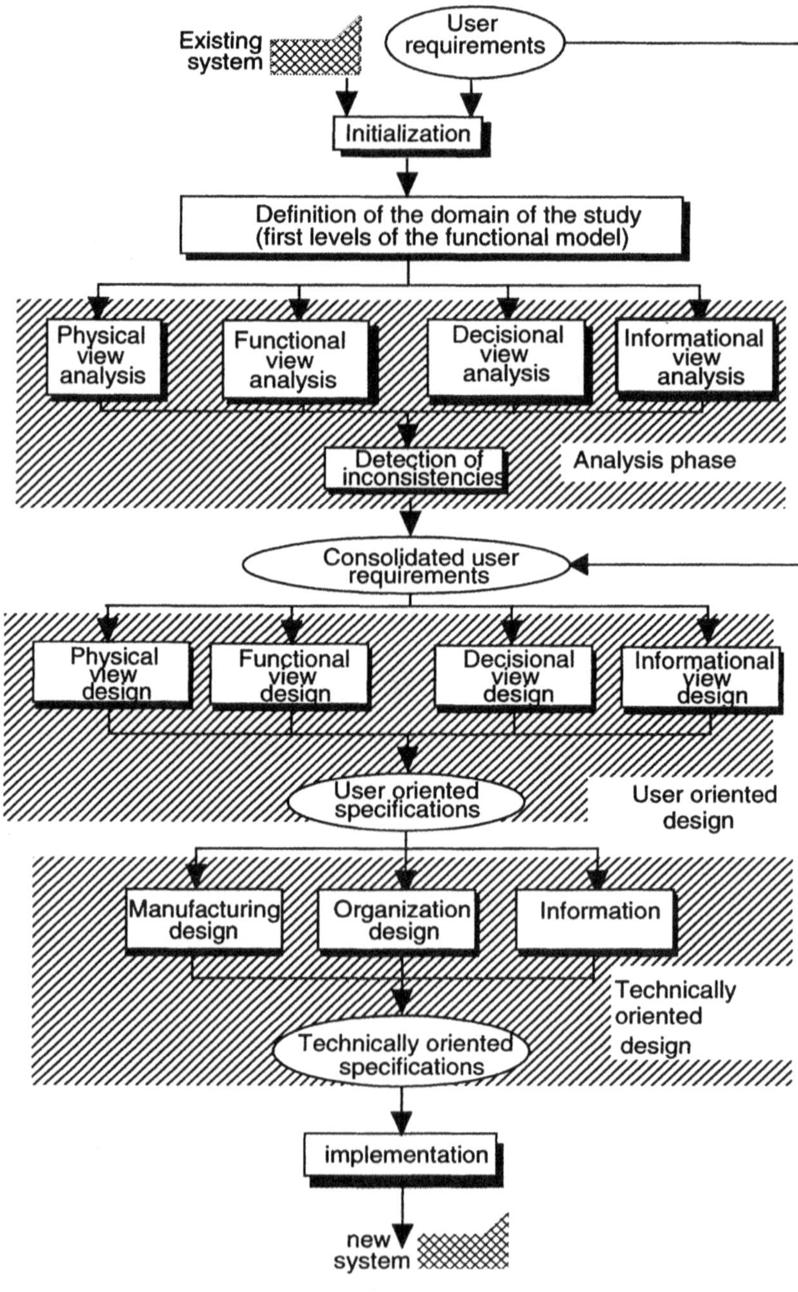

Figure 16-12 GRAI-GIM structured approach.

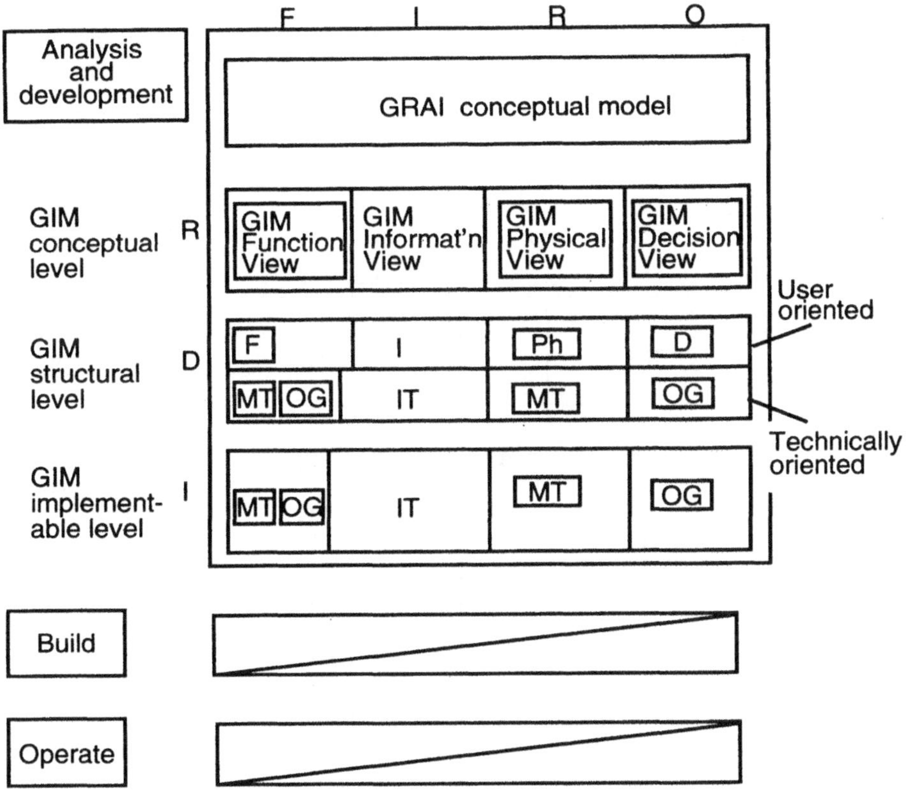

Figure 16-13 Mapping of GRAI-GIM onto the CIMOSA basic skeleton diagram see Fig. 16.9 for symbology used.

even though some of the figures have already been shown for comparison purposes in the earlier text of this chapter. The GRAI-developed text follows.

16.5 INTRODUCTION TO THE GRAI-GIM MAPPING

The objective of this section is to show the mapping of the GRAI-GIM (GRAI Integrated Method) Architecture onto the CIMOSA Architecture and the Purdue Enterprise Reference Architecture. The first part recalls some basic definitions on CIM, architecture and method. The second part gives the mapping of the GRAI-GIM architecture onto the CIMOSA architecture skeleton. The third part, which was written by J. Shewchuk during his stay

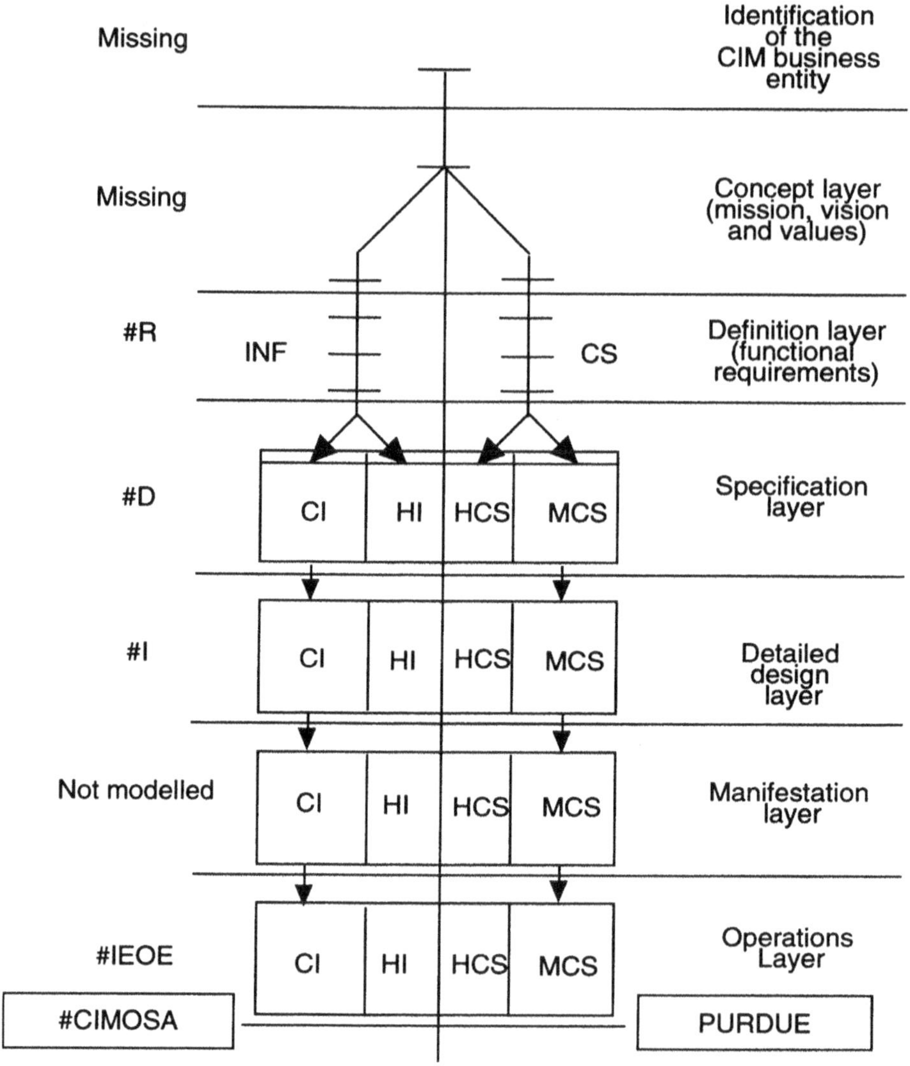

Figure 16-14 Mapping of CIMOSA capabilities on the structure of PERA.

in GRAI, shows the mapping of the GRAI-GIM architecture onto the Purdue architecture, as well as the mapping of the GRAI-GIM structured approach onto the Purdue approach.

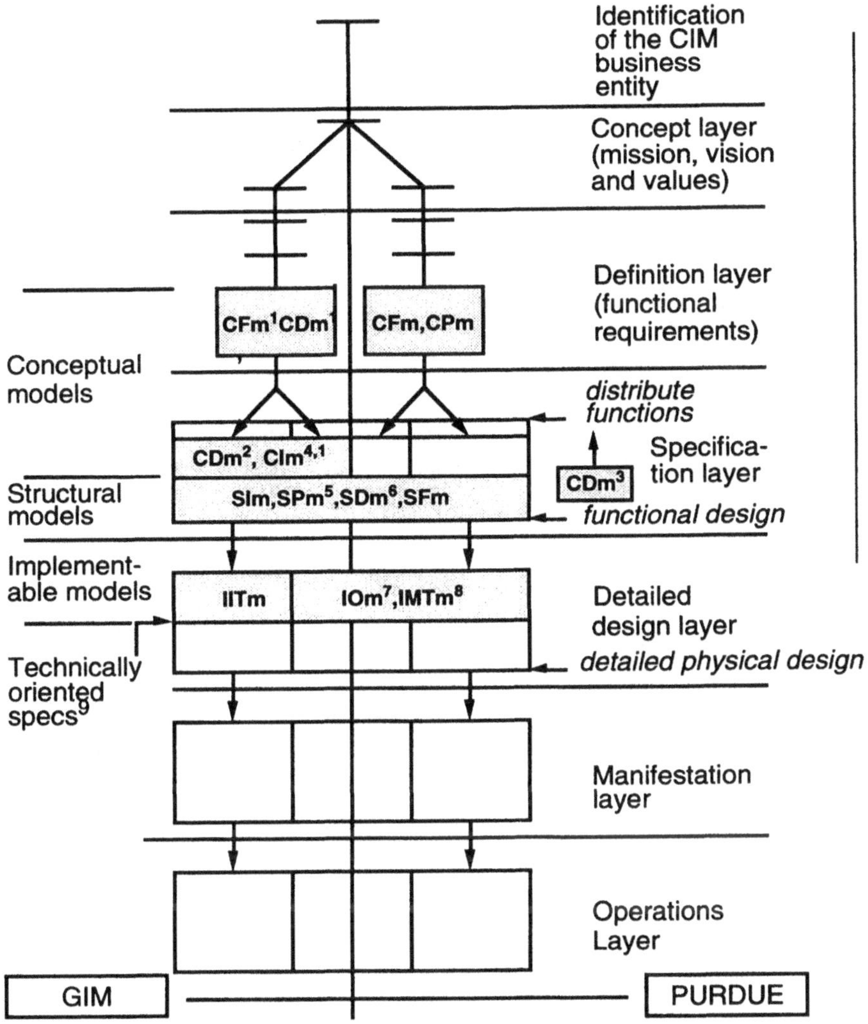

Figure 16-15 Mapping of GRAI-GIM modelling framework onto PERA (see Section 16.8.4 and Figures 16-21 and 16.22 for the explanation of the superscripts and abbreviations on the figure).

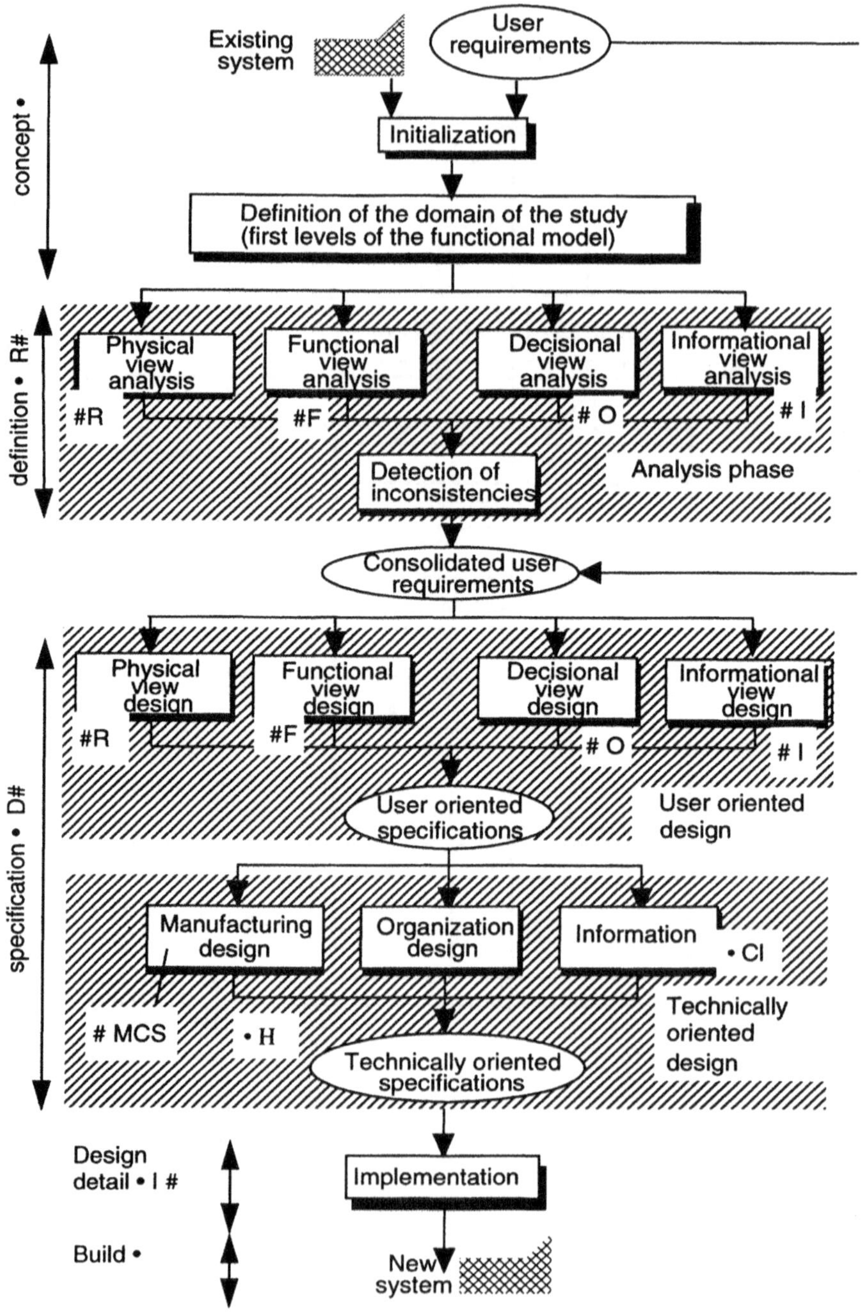

Figure 16-16 Mapping of PERA and CIMOSA on GRAI-GIM.

16.6 DEFINITIONS

16.6.1 What is CIM?

CIM stands for a global methodological approach to factory integration. It applies to all production operations from design to delivery. The integrated vision of CIM leads to the use of methods and requires the use of a mix of manufacturing techniques, information technology, computer science and automatic control theory. These methods and means allow for simultaneously increasing productivity, decreasing costs, fulfilling of due dates, achieving of quality standards, and for global and local flexibility. CIM requires the motivation and involvement of all people acting in the firm [1].

16.6.2 What is an architecture?

An architecture is a structured plan that can be seen as the basis for the definition of a product or an organization. This plan contains the rules and definitions of the properties of all constituent parts. With the rules and the defined properties, the parts are linked and result in a 'whole'. The plan contains all interrelationships between these parts and the information that has to be interchanged (via well defined interfaces) to enable all parts to work as one [2].

The architecture of a manufacturing system consists of a set of 'models' which represent various aspects of the total CIM system itself. The architecture can be seen as the basis for the design and implementation of a CIM system. These models, containing the elements and relationships between these elements, describe respectively the 'WHAT' (what CIM is conceptually composed of), the 'HOW' (how CIM works technically), and shows the way to transform the models into reality, e.g. working systems [1].

16.6.3 What is a method?

A method is a way of expressing or doing something according to a set of principles and an established sequence. Each method has its own application field; the area in which it is efficiently applied. First, a method should be composed of a philosophy which is a conceptual approach to the problem to be solved. The philosophy is often expressed by a conceptual model. A description is said to be at the conceptual level if it is describing a real world system (not necessarily realized yet, but perceived as a feasible system) in formalized abstract terms. Second, a method should be supported by tools or formalisms. By using these tools, several models of the studied system can be established. Third, a method should be a structured approach which gives a step-by-step guide for applying the method in an organized and efficient way [1].

16.7 MAPPING OF GRAI-GIM ONTO CIMOSA

16.7.1 Review of the GRAI conceptual model

(See also Chapter 7). The GRAI conceptual model decomposes a manufacturing system into three subsystems; decisional, informational, and physical, as shown in Fig. 16.17 (same as Fig. 7.6). The objective of this model is to identify and to structure a set of generic concepts for the modelling of any manufacturing system.

A modelling framework is defined (see Fig. 16.18) to categorize the kinds of models which need to be produced.

16.7.2 GRAI-GIM architecture framework

According to our current understanding of an architecture, it is possible to define three types of architectures representing respectively the 'what' and the 'how' of CIM systems, and the final definitive description of that system; the conceptual architecture, the structural architecture, and the implementable architecture.

The conceptual level defines the 'what', e.g. what functions, what physical resources, what decisions and what information are needed in a CIM system. It is independent of organizations, techniques, and enterprise constraints.

The structural level structures the elements defined in the conceptual architecture, by taking into account the various criteria and constraints of the structure. It shows how to distribute functions, information, decisions, and physical resources. The structural level is divided into two sublevels: a user oriented structural level which has the same views as the conceptual level, and a technical oriented structural level which has three domains.

The implementable level describes the final definition of a CIM system, by providing the 'picture' of the real system to be implemented. It repre-

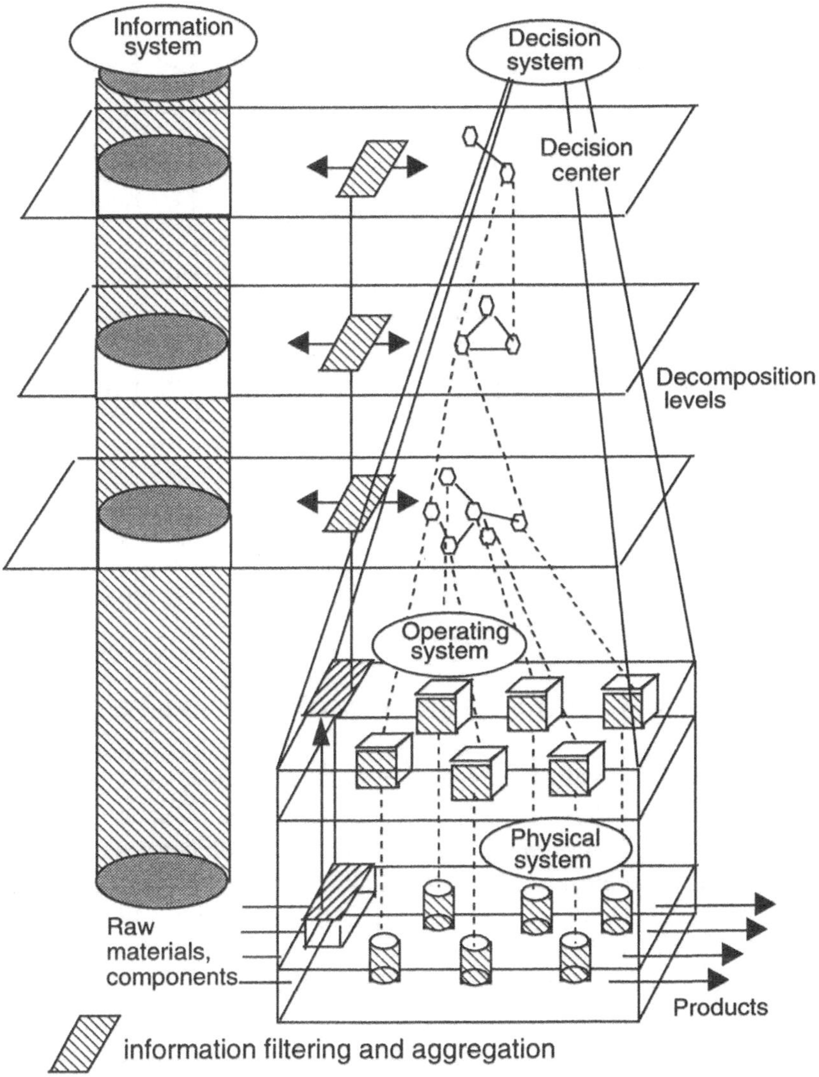

Figure 16-17 The GRAI conceptual model.

sents that system in terms of tangible information technology components, manufacturing technology components, and an organizational description.

The function view describes the decisional subsystem from the functional aspect. The decisional view models the decision subsystem. The

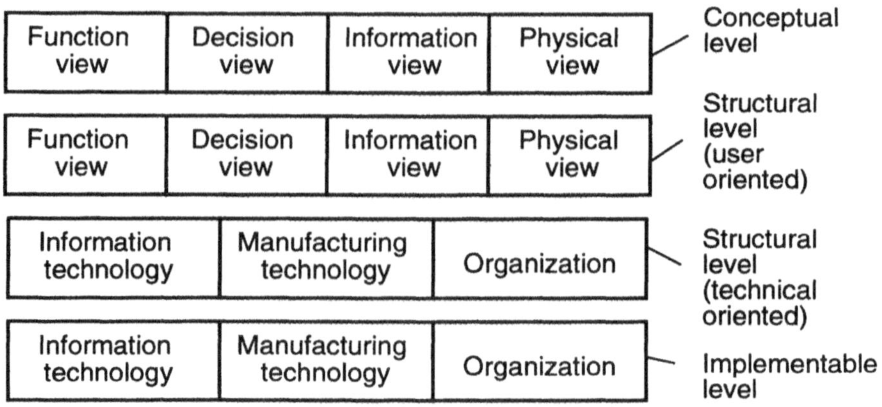

Figure 16-18 Framework of GRAI-GIM architecture.

information view models the information subsystem. The physical view is concerned with the physical subsystem.

16.7.3 Mapping of GRAI-GIM onto CIMOSA (see also Chapter 6)

16.7.3.1 Comparison of objectives and basic concepts

Both CIMOSA and GRAI-GIM apply to the CIM area. Due to the lack of a generally accepted definition for CIM, the respective objectives of CIMOSA and GRAI-GIM are closely related to each organization's own understanding of what CIM is, what problems are identified within the CIM area and to the way that CIM itself can be achieved.

The objective of CIMOSA is to develop a CIM reference architecture. It will be adopted by a user company to build its own CIM architecture which allows various 'Islands' to be formed and which ensures a good integration of all the CIM components. Particularly, CIMOSA insists that the released model of the CIMOSA architecture should be executable and evolutionary, e.g., it can be modified easily during the run time. CIMOSA also intends to provide a methodology which shows how to use the reference architecture in order to get a particular model. This is known as the CIMOSA model creation process, and as the CIMOSA System life cycle. Another objective of CIMOSA is to have CIMOSA standardized and accepted by both the CIM supplier community and the CIM user community. A standard CIM architecture will guide CIM vendors and suppliers in developing marketable products which are directly implementable in CIM systems designed according to CIMOSA. This means that integration can be ensured a priori if all the components to be implemented are developed according to CIMOSA requirements.

The objective of GRAI-GIM is to provide a methodology which shows how to transform a particular user's initial requirements into a set of detailed and precise technical specifications which will allow them to choose CIM components (or to develop) and to implement the new system. This methodology should define how to use reference architectures to build the particular architecture of a studied enterprise in order to save time, reduce cost and ensure a better integration a priori. The proposed GRAI-GIM intends to permit a CIM designer to pass directly from an initial set of user requirements to the development of the technical solutions (machines, softwares).

Therefore GRAI-GIM intends to provide at least two methods: one is a user oriented method allowing the transformation of initial user requirements into user oriented specifications for the business user's validation; another is a technically oriented method allowing the transformation of a user's specifications into a set of technical specifications. The technical specifications should contain all the detailed description needed to choose the manufacturing technology components, the information technology components and to define the organization of these components. Another objective of GRAI-GIM is to insist on the importance of the decisional aspect of a manufacturing system. This is because it is believed that one of the major deficiencies in available production planning and control systems is that they are too oriented towards the information processing rather than the decision making aspects of the overall problem. Decisional aspects are rarely taken into account and we do not yet know how to decentralize decision making.

GRAI-GIM considers that CIM is not confined to the technical aspect, but includes also the social, economic and human aspects. CIMOSA is more technically oriented although it does not consider that these other aspects are unimportant. Although no generally accepted definition for a CIM architecture and methodology exists, it is generally believed that the concept of an architecture and the concept of a methodology are different. An architecture provides a basic plan, a set of generic building blocks, models and rules, that allows one to describe the various aspects of a CIM system. A methodology (considered as a set of methods) provides guidelines and a structured approach which allow the designer to transform a set of user requirements into all the necessary technical specifications for implementing the system.

Neither CIMOSA nor GRAI-GIM are purely an architectural or a methodological approach. CIMOSA is more architecturally oriented but a methodology will also be developed. GRAI-GIM is more methodology oriented. In GRAI-GIM a reference architecture is considered as a support to the effort. GRAI-GIM is based on the GRAI method in that the decisional aspect is the most important aspect. CIMOSA is more functional analysis based. The analysis of behavior (i.e. the control structure, which is fed by the results of decisions) is a major aspect of CIMOSA. This key aspect will later allow the execution of models.

GRAI-GIM insists on the involvement and mobilization of people in every stage of the CIM system life cycle in order to capture the necessary knowledge. CIMOSA insists more on a technical procedure. Both CIMOSA and GRAI-GIM intend to achieve integration. GRAI-GIM insists on the consistency between the global objectives and the objectives of each element, and on the integration of the various systems which make up manufacturing. Although it is not as obvious as in GRAI-GIM, CIMOSA proceeds with the decomposition of the global objectives into a set of lower level ones. This is supported by the Function–Behavior–Structure methodology in the function view. But CIMOSA does insist on the role of 'Information Technology' in achieving integration.

Both CIMOSA and GRAI-GIM intend to provide aids for CIM design and implementation. CIMOSA insists on the instantiation process from a reference architecture to a particular architecture taking into account user requirements and constraints. GRAI-GIM insists on the transformation process from user requirements to a set of technical specifications taking into account the use of reference architectures.

While AMICE intends to make CIMOSA a standardized architecture, GRAI-GIM is a global methodological approach that cannot be standardized. GRAI-GIM considers that, 'CIM systems cannot be bought' they must be designed by the users themselves, who will choose and integrate various components – if possible, already available on the market – that they need.

16.7.3.2 Mapping of the GRAI-GIM architecture framework onto the CIMOSA skeleton

Fig. 16.19 shows a mapping of the GRAI-GIM architectural framework onto the CIMOSA architecture skeleton. The GRAI conceptual model corresponds to the Concept Layers of the CIMOSA skeleton. The conceptual level in GRAI-GIM can be compared with the Requirements Definition Modelling level of CIMOSA. The structural level of GRAI-GIM can be mapped onto the Design Specification Modelling level of CIMOSA.

However, it is important to note that the GRAI-GIM structural level has two sublevels: a User oriented structure and a Technical oriented structure. The User Oriented structural sublevel has four views identical to the four views of the Conceptual level. The Technically oriented structural sublevel has only three domains: an Information Technology Domain in which a set of technical specifications are elaborated to allow one to specify Information Technology components (computers, databases, LANs, softwares, etc.), a Manufacturing Technology domain for Manufacturing Technology components (machines, robots, transport means, human beings, etc.), and an Organizational Domain to specify an overall system organization. The Implementable level of GRAI-GIM can be compared to the CIMOSA Implementation Description Level. However, the difference is that the CIMOSA Implementation Description Level intends to generate a model which is executable, e.g., processable by the Services of the CIMOSA Integration Infrastructure, while the GRAI-GIM Implementable level intends to generate a description of how all the system components specified previously are installed physically in a real environment.

Some additional important comments need to be made concerning the details of the mapping of GRAI-GIM into CIMOSA as shown by Fig. 16.19.

The GIM physical view should be mapped onto the Resource View of CIMOSA because the goals and scope of these two are very close. However, the GRAI-GIM physical view covers only the resources of the physical subsystem (raw material transformation processes), while the CIMOSA resource view covers the resources of the whole manufacturing system. Therefore the box symbol used means, 'covers only a part of'.

The GRAI-GIM function view is not as complete as the CIMOSA function view. The GRAI-GIM function view usually stops at a high decomposition level, while the CIMOSA function view can reach to a detailed level such as the basic enterprise functions (Enterprise Activity) and Functional Operations. Again the box symbol is necessary.

The GRAI-GIM structural level has two sublevels (User Oriented and Technically Oriented) which could be mapped onto the CIMOSA Design Specification Level. The Information Technology Domain and Manufacturing Technology Domain don't belong to the User Oriented Structural Level.

The Organization Domain of GRAI-GIM cannot be mapped onto the CIMOSA Resource View. Likewise the Information Technology Domain of GRAI-GIM cannot cover the CIMOSA Organization view. And the Manufacturing Technology Domain of GRAI-GIM does not cover the CIMOSA Organization View.

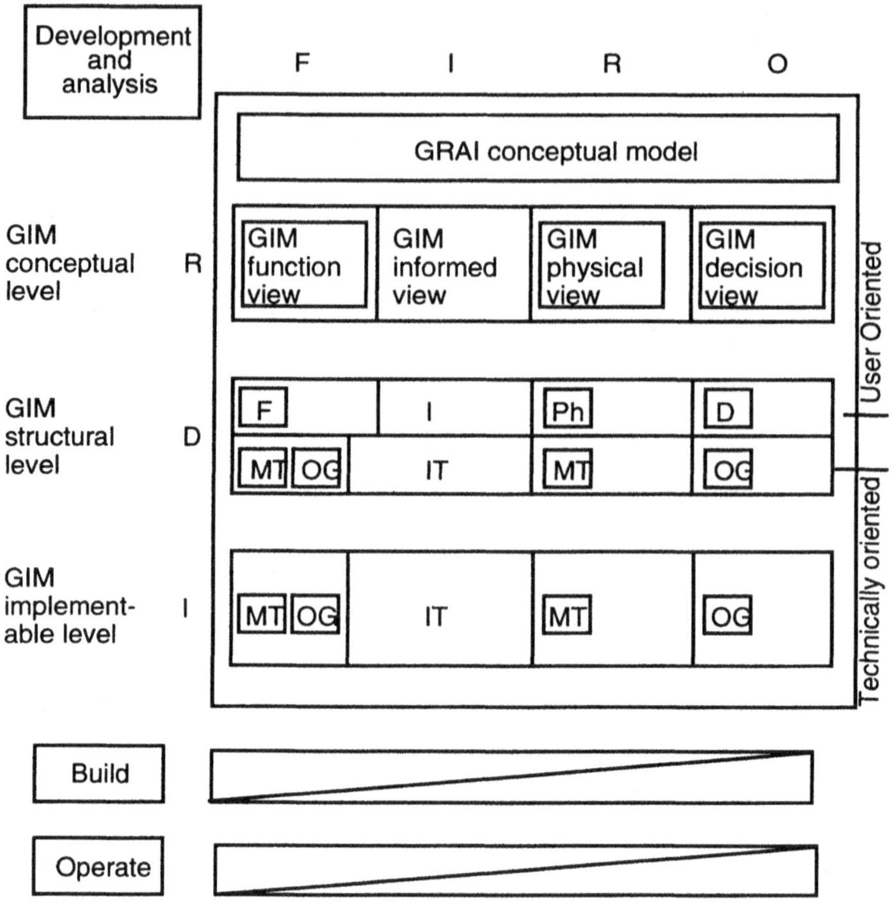

Figure 16-19 Mapping of GRAI-GIM onto the CIMOSA basic skeleton diagram Abbreviations: F: GRAI-GIM Function View : GRAI-GIM Information View Ph: GRAI-GIM Physical View D: GRAI-GIM Decision View IT: GRAI-GIM Information Technology Domain MT: GRAI-GIM Manufacturing Technology Domain OG: GIM Organizational Domain.

As noted, the small box means 'covers a part of'. For example, at the Requirements Definition Level of CIMOSA, the GRAI-GIM Function View cover only part of the CIMOSA Function View.

16.8 MAPPING OF GRAI-GIM ONTO THE PURDUE ENTERPRISE REFERENCE ARCHITECTURE

16.8.1 Introduction to GRAI-GIM to Purdue mapping

To compare the GRAI Integrated Methodology (GRAI-GIM) to the Purdue Enterprise Reference Architecture, it is desirable to map one architecture onto the other. However, we cannot simply 'map' GRAI-GIM onto the Purdue Enterprise Reference Architecture, as these are not the same types of entities. GIM is a methodology (containing an architecture) while the Purdue Enterprise Reference Architecture is just that – an architecture. The latter, however, is supported by several other documents which, taken together, comprise the Purdue Methodology for CIM. With these facts in mind, we must determine exactly what it is we wish to compare. The first step in doing this is to specify exactly the intentions, scope, and components of both the GRAI Integrated Methodology and the Purdue Enterprise Reference Architecture.

GRAI-GIM consists of a reference model, a modelling framework, a set of modelling formalisms, and a structured approach. The reference model describes the invariant parts of a CIM system; the subsystems, their relationships, and their behavior. The modelling framework shows the models which are used to represent the system and the relationship of these models to one another, whereas the modelling formalisms define ways in which specific models can be articulated, constructed, and manipulated. Finally, the structured approach shows the steps which must be followed to develop a CIM system, where the system is defined by a reference model. The scope of GRAI-GIM is the design and specification of CIM systems. The specification of the generic functional tasks, data elements and relations, etc. which are required by all CIM systems, is outside the scope of GRAI-GIM, but such information has been developed within the scope of other projects (e.g. ESPRIT Project 38 – IMPACS) using GRAI-GIM.

The Purdue Enterprise Reference Architecture [3] is itself only an architectural specification. It has developed in conjunction with two other documents, however, which together with the Purdue Enterprise Reference Architecture comprise the 'Purdue Methodology for CIM'. These documents are the 'Purdue Reference Model' [4] and the 'Purdue Implementation Procedures Manual for Developing Master Plans for CIM' [5]. The former specifies, amongst other things, the specification of the generic tasks which are required by all CIM systems, while the latter specifies the steps which must be followed to develop a CIM system. The Purdue Enterprise Reference Architecture is very broad in scope, being intended for any type of manufacturing or service enterprise.

Based upon the above, the following can be compared between GRAI-GIM and the Purdue Methodology:

1. Definition of CIM systems; the subsystems, their relationships, and their behavior. In GRAI-GIM, this is explicitly defined as the 'reference model of a CIM system'. In the Purdue methodology, these concepts are presented as the 'basic concepts' upon which the Purdue Enterprise Reference Architecture is based (this match is not perfect, however).
2. Modelling frameworks. The GRAI-GIM modelling framework, where the modelling approach is considered, shows all the models used and their relationships with one another. This can be compared to the Purdue Enterprise Reference Architecture, where the various subarchitectures and their relationships are shown.
3. Modelling formalisms. In GRAI-GIM, a formalism is specified for each model, and there is a very close link between modelling formalisms and models (especially so for the decisional models). GRAI has developed the GRAI grids, GRAI nets, and GRAICO formalisms for use with GRAI-GIM. In the Purdue Enterprise Reference Architecture, the emphasis is more on specifying the requirements of the models and suggesting formalisms which may be used. Purdue has developed a 'modular task and functional representation method' which can be used for activity modelling.
4. Structured approach (implementation procedure). The GRAI-GIM structured approach serves the same function as the Purdue Implementation Procedures Manual for Developing Master Plans for CIM, but the scope of the former is not as large. GRAI-GIM concentrates on system analysis and design, and ends with the development of 'technical oriented specifications', which may be used to purchase or otherwise obtain the required resources. The Purdue Implementation Procedures Manual for Developing Master Plans for CIM covers all steps from system analysis and design through construction of and installation of equipment in the new plant plus plant operation itself.

Each of these, except for modelling formalisms, shall be discussed in this section to varying extents. For the first item, we shall simply point out the concepts which are important for understanding the differences between GRAI-GIM and the Purdue Enterprise Reference Architecture. This will set the stage for focusing on the mapping of the GRAI-GIM modelling framework onto the Purdue Enterprise Reference Architecture.

16.8.2 Basic concepts for architectural comparison

The following concepts are important for understanding the differences between GRAI-GIM and the Purdue Enterprise Reference Architecture.

16.8.2.1 GRAI-GIM

Three types of activities exist in manufacturing systems:

1. *Decisional*; itself subdivided into two further types of decisional activities
   (a) 'To Do (or 'To Execute'): an activity which 'systematically' transforms an initial state to a resultant state, i.e. given an initial state, we can execute the activity according to a well defined algorithm. These are all automatable.
   (b) 'To Decide': a decisional activity which transforms an initial state into a 'nonsystematic' resultant state, according to specific criteria (objectives, constraints, criteria, variables). They comprise the human based, innovative functions.
2. *Informational*; activities which concern transmitting, processing, and memorizing the information needed by decisional activities.
3. *Physical*; physical activities aimed at transforming components into final products, adding value along the way. These activities are 'controlled ' by the decisional activities.

Based upon these three types of activities, there are three systems in all manufacturing enterprises: (1) a decisional system, (2) a informational system, and (3) a physical system. The interface between the decisional and physical system is known as the operating system.

Two types of specifications are required for analyzing and designing CIM systems:

1. *User oriented*: defines what must be done, from the user's point of view.
2. *Technically oriented*: defines the components required from the point of view of technology, i.e. in a language understandable to component vendors or developers.

16.8.2.2 Purdue Enterprise Reference Architecture

There are two types (streams) of transformations (activities) in any enterprise:

1. *Informational transformations*: decision making, control, communications, information storage.
2. *Physical or Manufacturing transformations*: material or energy conversion, movement, storage.

All activities can be classified in terms of whether they can, cannot, or must be automated (Table 16.2 at the end of this chapter). The extent to which these activities are distributed amongst the three groups defines the levels of *automatability* and *humanizability* which are possible for the system(s) which must execute the activities (Fig. 16.20) (same as Fig. 8.7).

All transformations can be defined in a common manner; they transform inputs to outputs, are executed with transformation mechanisms, and require enabling parameters to be executed. The question of automatability does not change the way tasks are modelled – automatability is an implementation concern (e.g. execute an automatable informational task with an algorithm; execute a nonautomatable informational task with an algorithm; and a nonautomatable informational task with a written statement of requirements, objectives, constraints, etc. for a human).

Based on the concepts of task type/degree of automatability used for executing the tasks, the actual enterprise consist of three subsystems:

1. An Information Component: for executing the informational transformation activities performed which are automated.
2. A Manufacturing Component: for executing the physical (manufacturing) transformation activities which are automated.
3. A Human and Organizational Component: for executing the informational and manufacturing transformation activities which are performed by humans.

The relative size of the Human and Organizational Component is defined by the extent of automation chosen for the enterprise. This depends upon many factors, but will be bound by the limits of automatability and humanizability (Fig. 16.20).

Manufacturing systems are seen from two different points of view:

1. *The Business Environment* – users of the system.
2. *The Enterprise's Resources* – point of view of technology.

16.8.3 Architectural mapping

We shall briefly present each architecture before mapping GRAI-GIM onto the Purdue Enterprise Reference Architecture.

16.8.3.1 GIM modelling framework

The GRAI-GIM modelling framework has two dimensions: an abstraction level and a dimension of the system life cycle.

1. Abstraction Level – 3 levels used:

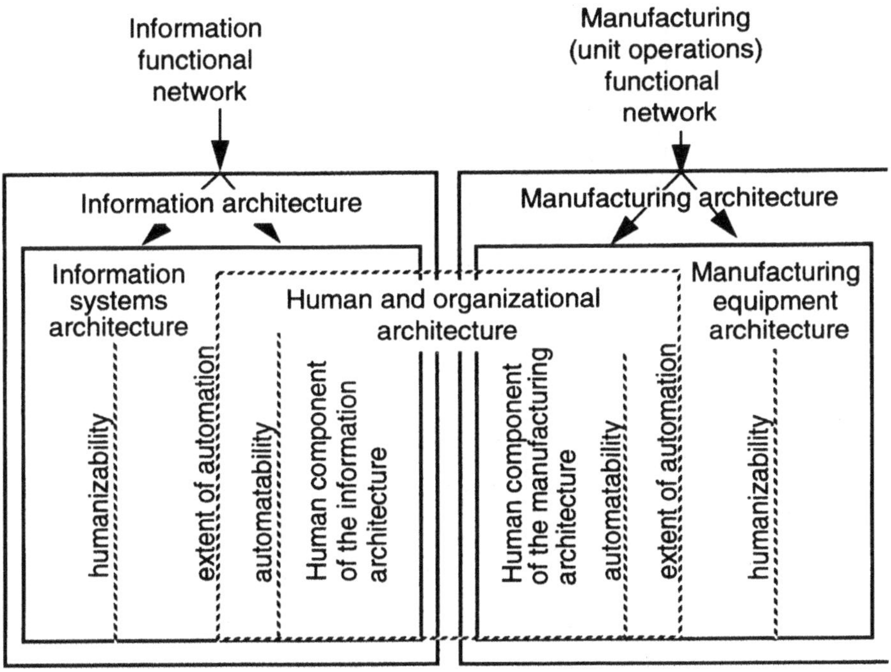

Figure 16-20 Task type and degree of automatability.

(a) *Conceptual*: defines what must be done, without regard to any organizational or technical concerns (i.e. by who, when, where, and how).

(b) *Structural*: defines who (what) executes what is specified at the conceptual level, and when and where.

(c) *Implementable*: defines how to execute what is defined at the conceptual levels, considering technical and organizational constraints.

2. Level of Functional Decomposition – results in different views of the system. Two types of functional decomposition, depending upon the approach used:
   (a) User oriented – informational, physical, decisional, and functional models. The first three result from the manufacturing system definition in the reference model.
   (b) Technically oriented – organizational, information technology, and manufacturing technology, corresponding to the three categories of the technically oriented specifications.

The GRAI-GIM architectural framework, showing the various models and the purpose of each model, is shown in Figures 16.21 and 22.

	Views				
	Informational	Physical	Decisional	Functional	
conceptual models	CIm – description of all stable and natural data of the organization; attributes and relationship between data	CPM – description of physical processing activities showing precedence and connectivity and product flows over activities	CDm – macro view, period-driven activities: group functional and temporal (H,P) criteria; identify and locate flows of information, decision variables, objectives and constraints. – macro view, event driven decisional activities: group by function and control hierarchy. – macro view, all decisional activities; identify initial and final states, triggers and information supports	CFM description of all activities in manufacturing enterprise, showing precedence and connectivity; identification of all functions without specifying where or how performed	useroriented models

Figure 16-21 GRAI-GIM architectural framework (conceptual models).

	Informational	Physical	Decisional	Functional	
srucutral models	SIm – description of how to organize and group data according to requirements of users.	SPm – description of structural requirements of physical processing system: number of processors (machines, human, etc), arrangement into cells or shops; who does what.	SDm – macro view: description of who is responsible for each group of decisional activities, how to group these groups into departments, etc. – micro view: description of all resource requirements for all decisional activities.	SFm – description of resource requirements for all activities: grouping of activities based upon various criteria (flexibility, economic factors, etc).	User oriented models

	SOm	SITm	SMTm	
implementational	IOm – organization of physical system (machines, material handling, people). – specification of working and training procedures, etc. – organization of mgmt structure. – specification of responsibilities, mgmt rules, etc.	IITm – hardware: computers and networks, file servers, data storage devices, etc. – software: generic tasks (operating system, database mgmt etc), application software (CAD, etc), decision support tools	IMTm – specification of all resources used at physical level: machines, material handling, people, etc – specification of all resources used for control and interface directly with physical resources (actuators, sensors, etc).	technically oriented models

Figure 16-22 GRAI-GIM architectural framework (structural and implementational models).

16.8.3.2 Purdue Enterprise Reference Architecture

The Purdue Enterprise Reference Architecture modelling framework is based upon two dimensions: view and activity classification.

1. View – Two views are used:
   (a) *Functional*; description of what has to be done, without regard to how, when, where, etc.
   (b) *Implementation*; defines how to attain what has been specified in the functional view; what resources are required, how they are arranged, etc.
2. Activity Classification – two classification schemes, depending upon view used:
   (a) *Functional*; informational and physical (manufacturing).
   (b) *Implementation*; informational transformation activities performed with automated systems, physical (manufacturing) transformation activities performed with automated systems, and informational and manufacturing transformation activities performed by humans.

The Purdue Enterprise Reference Architecture, showing the various models and the purpose of each model, is shown in Fig. 16.23.

16.8.4 Mapping of GRAI-GIM architectural framework onto the Purdue Enterprise Reference Architecture

The mapping of the GRAI-GIM modelling framework onto the Purdue Enterprise Reference Architecture is shown in Fig. 16.15. Note that only those Models present in the GRAI-GIM modelling framework are mapped here. The GRAI-GIM items which map against the top layers of the Purdue Architecture are part of the GRAI-GIM structured approach. The mapping of the GRAI-GIM Structured Approach against the Purdue Model is shown in the consecutive figures 16.24 a,b, and c. Particulars behind the mapping of the various models are described in the notes below (refer to Fig. 16.15):

1. Specification of initial and resultant states for decisional activities in the CIM Conceptual Decisional Model (micro view) is equivalent to the specification of conditions existing before and after informational (decisional) transformations occur in the Informational Functional Architecture of the Purdue Enterprise Reference Architecture.

   Similarly, specification of trigger supports for decisional activities is equivalent to the specification of conditions under which informational (decisional) transformations occur. The GIM Conceptual Decisional Model Macro view shows the identification and specification of flows (information, decision, variables, objectives, and constraints) between

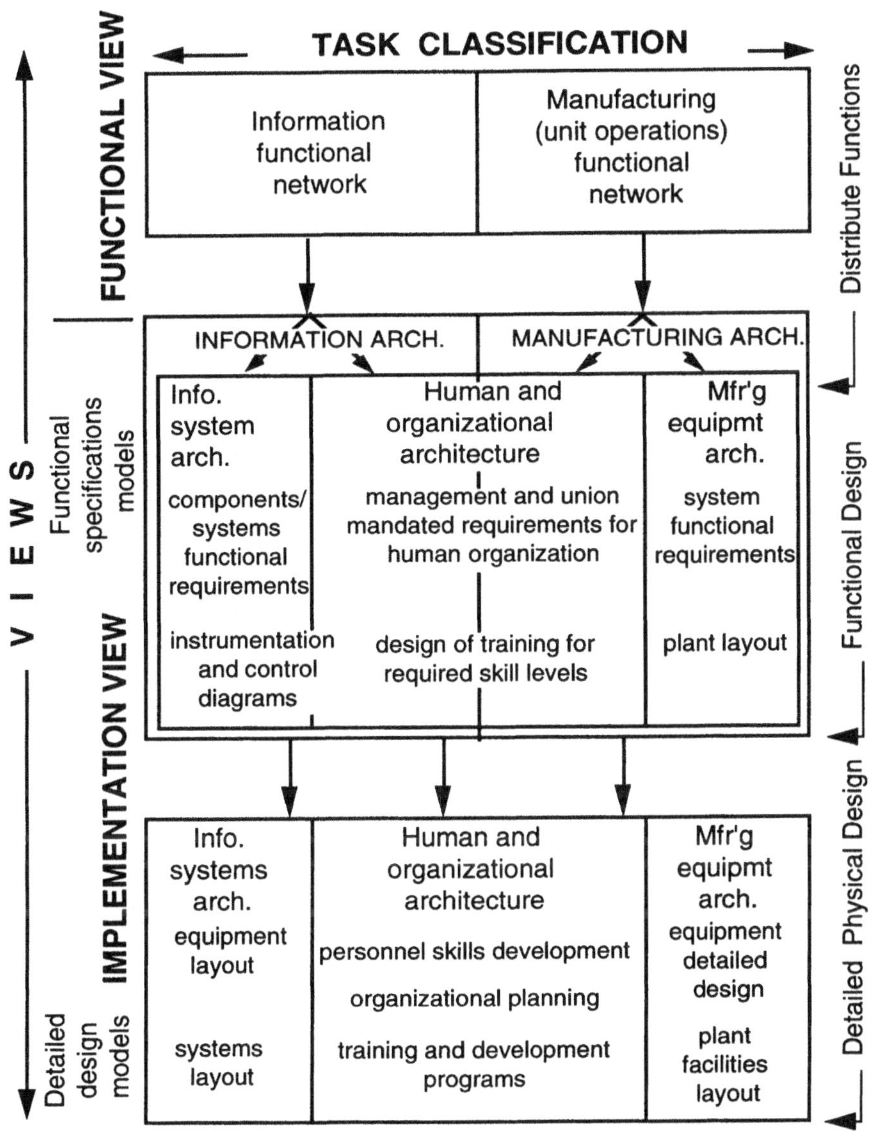

Figure 16-23 Purdue Enterprise Reference Architecture.

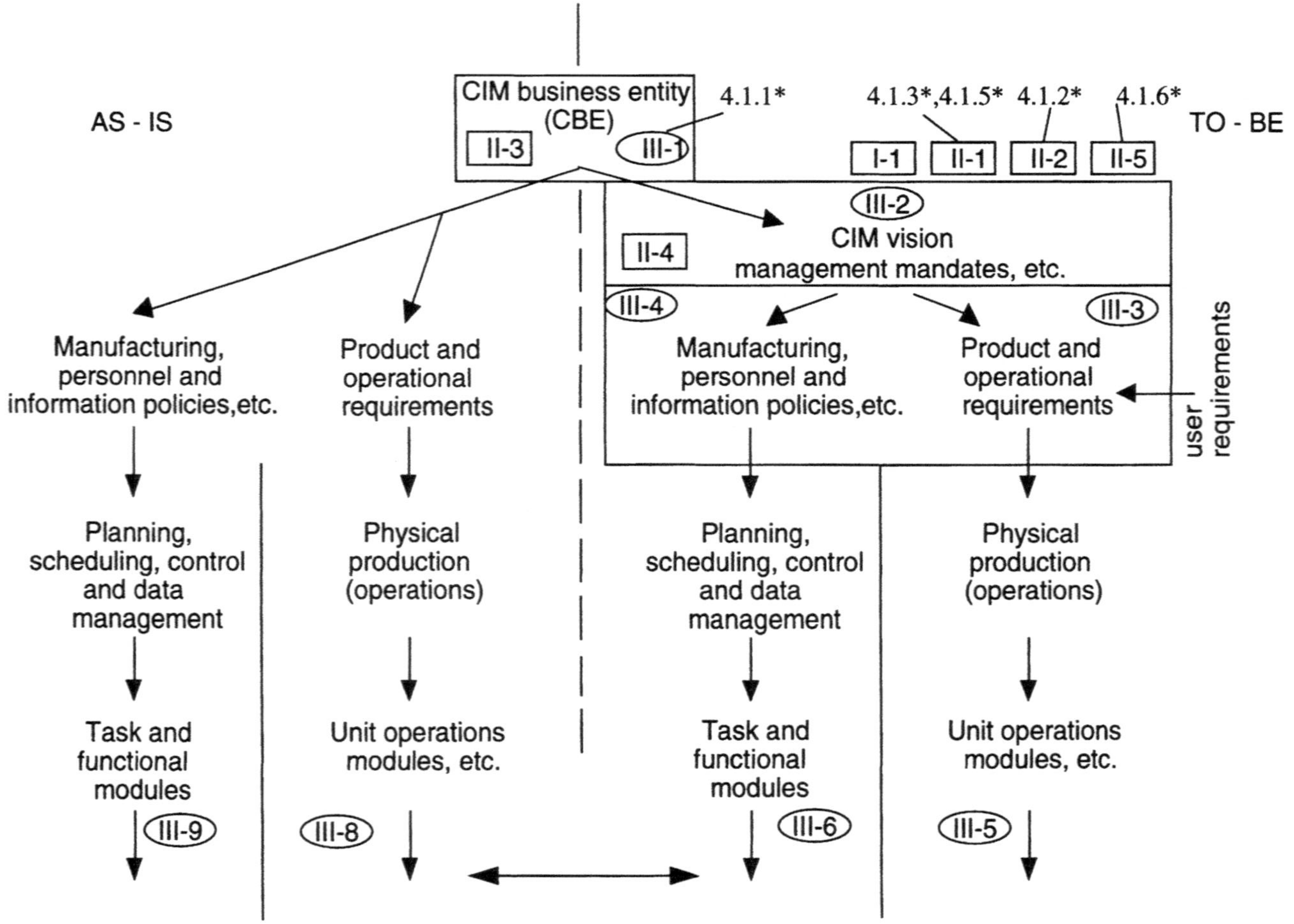

Figure 16-24 (a) Mapping of the GRAI-GIM structured approach onto the Purdue Implementation Procedures Manual. (* denotes GRAI-GIM structured approach initialization phase step number (see also section 16.8.5.1)

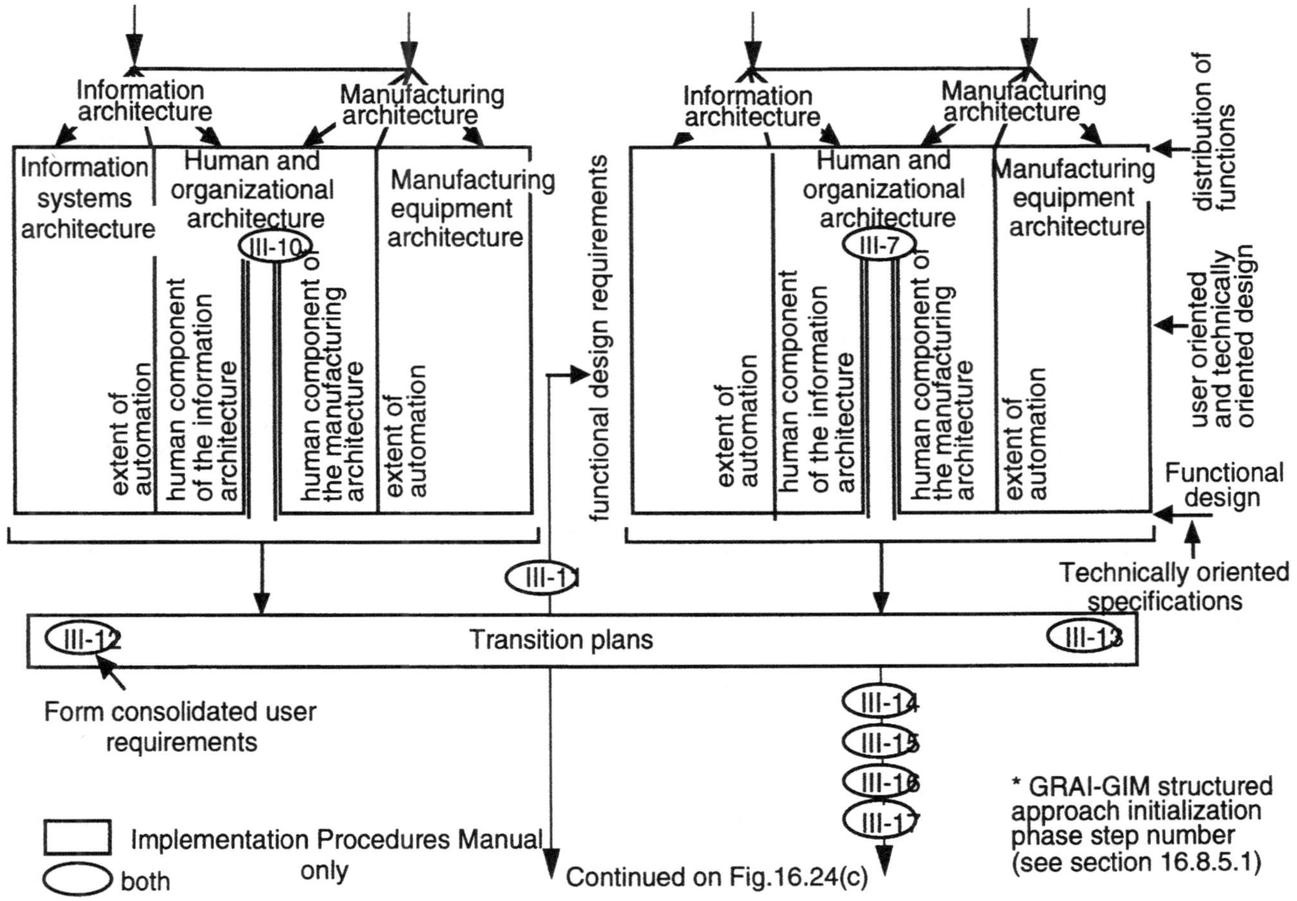

Figure 16.24 (b) Mapping of the GRAI-GIM structured approach onto the Purdue Implementation Procedures Manual (Continued).

decision making groups as well as how global objectives are decomposed into subobjectives for the groups. These can be performed in the Information Functional Architecture of the Purdue Enterprise Reference Architecture using a data flow model (e.g. data flow diagram) for the former, and any model showing hierarchical decomposition for the latter.

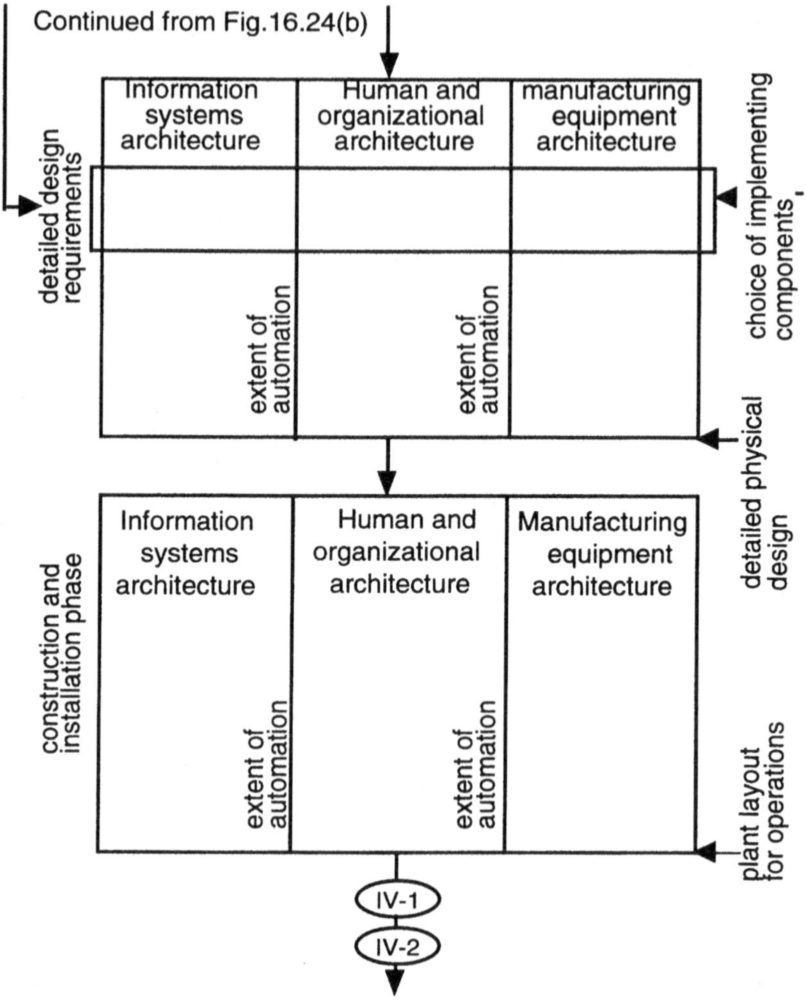

Figure 16.24 (c) Mapping of the GRAI-GIM structured approach onto the Purdue Implementation Procedures Manual (Continued).

2. Specification of informational supports for decisional activities in the GIM Conceptual Decisional Model (micro view) is equivalent to the specification of mechanism requirements for informational (decisional) transformations in the Information Architecture of the Purdue Enterprise Reference Architecture. In the GIM Conceptual Decisional Model macro view, decisional activities are grouped by functional and decisional criteria. At the higher levels of the decisional system, decisional activities are period driven and the decisional criteria is thus temporal (i.e. horizon and period), whereas at the lower levels decisional activities are event driven and the decisional criteria is thus control oriented (activities triggering other activities). A single model containing this same information is not presented in the Purdue Enterprise Reference Architecture. Grouping of all information activities by control hierarchy is performed when assigning activities to modules in the Purdue Scheduling and Control Hierarchy in the Information Architecture of the Purdue Enterprise Reference Architecture.
3. Identification of decisional activities as being of type 'To Execute' or of type 'To Decide' in the GIM Conceptual Decisional Model (micro view) is equivalent to the task of distributing informational (decisional) activities between the Information System Architectures and the Human and Organizational Architectures at the functional specification level in the Purdue Enterprise Reference Architecture.
4. The GIM Conceptual Information Model shows data elements, attributes, and relationships between elements. This is not a concern at the functional requirements level in the Purdue Enterprise Reference Architecture, as these items are not functional requirements themselves, but rather result from the specification of informational functional requirements. They become important only when trying to develop the systems for implementing the informational activities defined in the Information Functional Architecture. Thus, the GIM Conceptual Informational Model is constructed during development of the Information Architecture in the Purdue Enterprise Reference Architecture.
5. GRAI-GIM Structural Physical Model specifies, amongst other things, who does what. Physical activities which are performed automatically will be assigned to automatic control devices as specified in the Information System Architecture, and physical activities which are performed manually will be assigned to humans as specified in the Human and Organizational Architecture (assignments made at functional specification level).

6. A description of resource requirements for all decisional activities in the GIM Structural Decisional Model (micro view) is performed during development of the Information Systems Architecture (informational activities which are automated) and the Human and Organizational Architecture (informational activities which are performed by humans) in the Purdue Enterprise Reference Architecture (functional specification level). In the GIM Structural Decisional Model macro view, the description of who is responsible for each grouping of decisional activities, how to group these groupings into departments, etc. is equivalent to the assignment of control tasks to processors and grouping the processors during systems development. This occurs in the development of the Information Systems Architecture for automated control tasks, and in the Human and Organizational Architecture for human control tasks (functional specification level).
7. GRAI-GIM Implementable Organizational Model is concerned with specification of all resources used at the physical level, including both men and machines. In the Purdue Enterprise Reference Architecture, the former is covered by the Human and Organizational Architecture and the latter by the Manufacturing Equipment Architecture, both at the detailed design level.
8. GRAI-GIM Implementable Manufacturing Technology Model is concerned with the specification of all resources used at the physical level, including both men and machines. In the Purdue Enterprise Reference Architecture, the former is covered by the Human and Organizational Architecture and the latter by the Manufacturing Equipment Architecture, both at the detailed design level.
9. GIM technically oriented specifications are used for choosing in an 'integrated way', the components to use in the CIM system. In the Purdue approach, equipment is selected during detailed design and then used in development of the detailed physical design.

16.8.5 Comparison of the GRAI-GIM structured approach to the Purdue Implementation Procedures Manual

16.8.5.1 GRAI-GIM

The GRAI-GIM structured approach is shown in Fig. 16.12. The initialization phase is characterized by the following steps (the numbering here is taken from Doumeingts [1]):

- Determination of the overall definition of the company (4.1.1).
- Presentation of the method to the company managers (4.1.2).

- Definition of the domain of study (4.1.3).
- Diagnosis to get another point of view on all problems (4.1.4).
- Initialization Milestone – update domain objectives, constraints, etc (4.1.5).
- Definition of the people involved in the study (4.1.6).
- Training of the participants (4.1.7).
- Planning of resources for the next phase (4.1.8).

16.8.5.2 Purdue Implementation Procedures Manual

Table 8.2 Chapter 8 shows the chapter title list from the Purdue Implementation Procedures Manual for Developing Master Plans for CIM. Fig. 16.24 a,b and c show how the material presented in these chapters maps onto the Purdue Enterprise Reference Architecture. The listing immediately above from Doumeingts [1] is used in the figure to represent the same topics.

16.8.5.3 Mapping of GIM Structured Approach onto the Purdue Implementation Procedures Manual

Fig. 16.24 also shows the mapping of the GIM Structured Approach onto a representation of the Purdue Enterprise Reference Architecture in both its To-Be and As-Is plants which shows also how the Implementation Procedures Manual relates to the architecture. The labels in the various areas of the architecture diagrams show where the corresponding manual chapters fit. A fundamental difference in the approaches is the point at which the existing system is analyzed. In the Purdue approach, the objectives, goals, etc. are first defined, then the new system is functionally defined, then the existing system is analyzed and defined. The objective is then to determine how to go from the existing ('As-Is') system to the new ('To Be') system. In the GRAI-GIM structured approach, the objectives, goals, etc. (user requirements) are first defined, then the existing system is analyzed, and then the new system is designed, based upon the user requirements and analysis of the existing system.

16.9 REFERENCES

[1] Doumeingts, G., et al, GRAI-GIM Integrated Methodology, Version 1.0, LAP/GRAI Laboratory, University of Bordeaux 1, Bordeaux, France (May 1992).

[2] ESPRIT Consortium AMICE, Open System Architectures, CIMOSA, ADIO, Architecture Description, ESPRIT Consortium AMICE, Brussels, Belgium (1991).
[3] Williams, T.J., The Purdue Enterprise Reference Architecture, Instrument Society of America, Research Triangle Park, NC (1992).
[4] Williams, T.J., Editor, A Reference Model for Computer Integrated Manufacturing (CIM), A Description From the Viewpoint of Industrial Automation, Minutes, CIM Reference Model Committee, International Purdue Workshop on Industrial Computer Systems, Purdue University, West Lafayette, IN (May 1988); Instrument Society of America, Research Triangle Park, NC (1989).
[5] Industry–University Consortium, Purdue Laboratory for Applied Industrial Control, An Implementation Procedures Manual for Developing Master Plans for Computer Integrated Manufacturing, Report Number 155, Williams, T.J., Editor, Purdue Laboratory for Applied Industrial Control, Purdue University, West Lafayette, IN (June 1992).

Table 16.2 Task type and degree of automatability

Degree of automatability	*Informational transformations*	*Physical transformations*
Tasks which may be automated	Task operation is totally definable in mathematical and / or computer programming terms	Task is within physical and technological capacities and capabilities of physical processing equipment
Tasks which can not be automated	Task requires human innovation / thought	Tash requires human capabilities or capacities (e.g. some type of assembly)
Tasks which must be automated	Task requirements beyond human capacities and / or capabilities (e.g. processing speed or accuracy requirements)	Task requirements beyond human capacities and / or capabilities (e.g. strength, endurance)

17

Mappings against a matrix

P. Bernus, L .Nemes and T.J. Williams

17.1 THE MATRIX OF REQUIREMENTS OR NEEDS

As noted in the previous chapter, it was the goal of the Task Force to develop an evaluation method for the candidate architectures which considered them on the basis of a set of general requirements for these architectures as a set of user's needs.

At the Australia Section Meeting in Melbourne, it was decided to try to develop such a method. We chose to use as a basis the major differentiating aspects of each architecture's modelling framework or structural diagram. These were as follows:

- The Generic, Partial and Particular separation of CIMOSA in developing architectural models.
- The Function, Information, Decisional/Organization and Structural Views of GRAI-GIM.
- The Information and Customer Service and the Human and Machine separation categories from the Purdue Architecture.
- The Identification, Concept, Requirements, Design, Implementation, Build and Operate Phases as a composite of the Life History presentation of each of the architectures.

In the next section, we discuss the reasons for the choice of the above criteria of evaluation and the expected results. We also show the development of the basic format as it evolved from discussions with the other members of the Task Force.

17.2 RATIONALE OF THE MATRIX PRESENTATION

17.2.1 The goal

We are developing here a matrix presentation form which will be suitable for the characterization of generic enterprise architecture proposals. It will be used for the following purposes:
1. Comparison of existing architectures
2. Pointing out areas not covered in any of the architectures

17.2.2 The difficulty

Different architectures use different ways to divide concerns. For example, one architecture may divide the manufacturing system into control system and process, another may divide it into subprocesses each with the process and control packed together, etc.

In order to discover the important correspondences between the architectures we need to look at the type and the structure of the descriptions proposed by the candidate architectures and point out the relationships between the constituent types of information recorded in these descriptions.

17.2.3 The agenda

- Identify how these divisions relate to one another
- Present a matrix which can accommodate the scope covered by the models of any of the investigated architectures
- Fill in a matrix for each architecture
- Draw conclusions

The resulting matrix will be similar to the 'CIMOSA cube' but:
- Descriptions, not necessarily models will be evaluated (proposals widely differ in the degree of formality of presentation).
- We intend to show how much detail has actually been worked out by each development group.

In terms of the increasing degree of detail included by each architecture the following degrees of presentation will be indicated:
- Only the category of description has been defined, no generic description (model) is provided which could be standardized or reused.
- The above plus an extensive set of guidelines is made available to facilitate the generation of the necessary descriptions (models).

- The above plus generic models and examples (blueprints) are available. These can be reused as recipes for an 'enterprise cookbook' or could be standardized.

17.2.4 The dimensions of the matrix

17.2.4.1 An explanation of the rationale of the matrix

1. CIMOSA proposes three different properties as their basis for model categorization. These are:
   (a) By the role of model in the functional model of the derivation process (Requirements, Design, and Implementation),
   (b) By the place in the model instantiation process (Generic, Partial, Particular), and
   (c) By the aspect of modelling (Function, Information, Organization, Resource).
2. Purdue adds another three:
   (a) By the relationship to the enterprise mission: Control information vs. Customer service,
   (b) By the implementation means used: human vs. machine, and
   (c) By the temporal sequence as produced in the life-cycle: e.g., management mission and vision → management requirement → task specification → module requirements →functional design or specification → detailed design → construction and check out → operation.
3. GRAI-GIM defines its structure:
   (a) By the use of a different set of aspects of modelling (Function, Information, Decision, Physical structure).
   (b) By the use of a different set of roles in the model derivation process (Requirements, User specification, Technical specification).
4. The SATT methodology[1] defines two additional dimensions of modelling category not used in other modelling technologies or architectures:
   (a) By the type of the object being modelled (Life-cycle models vs. System models – grouped into evolution layer and system layer respectively). Evolution layer: how the system is conceived, built and modified. System layer: what the system does when implemented.

1. SATT is a methodology developed by Peter Bernus, Laszlo Nemes and others at the Computer and Automation Institute of the Hungarian Academy of Sciences the late 1970's and early 1980's. It was published in full only as proprietary material of the Institute and only in Hungarian. This technique was not evaluated by the Task Force in its studies.

(b) By the degree of abstraction of the model: (Abstract vs. Concrete) – i.e. functional model types vs. physical model types. Note: correspondingly on the functional level SATT talks about models while on the physical level SATT talks about modules).

5. Two additional natural divisions are also included:
   (a) Hardware vs. Software. The Hardware versus the Software categorization of SATT is considered important as an addition to the matrix evaluation method.
   (b) Human vs. Material processing machine vs. Information processing machine (computer), somewhat similar to that used by Purdue.

17.2.4.2 Dimensions defined as a result

The following domains are therefore given by the investigated architectures:

- (Requirements – Design – Implementation) [CIMOSA]
- .(Generic – Partial – Particular) [CIMOSA]
- (Control and Information – Customer service) [Purdue]
- (Human – Machine) [Purdue]
- (Temporal sequence of model production) [All]
- (Function – Information – Decision – Physical structure) [GRAI-GIM]
- (Requirements – User specification – Technical specifications) [GRAI-GIM]
- (Object is the project – Object is the system) [SATT]
- (Hardware – Software) [SATT]
- (Human – Material processing machine – Information processing machine) [SATT and Purdue]

Out of the above list item 4 is a subpart of item 10 and items 5 and 7 are not independent model categories, rather they define clusters of models and not independent dimensions. Therefore the dimensions of the resulting matrix after the first iteration are:

- (Requirements – Design – Implementation) [CIMOSA]
- (Generic – Partial – Particular) [CIMOSA]
- (Control and Information – Customer service) [Purdue]
- (Function – Information – Decision – Physical structure) [GRAI-GIM]
- (Object is the project – Object is the system) [SATT]
- (Hardware – Software) [SATT]
- (Human – Material processing machine – Information processing machine) [SATT and Purdue]

Still, 864 combinations are conceivable (c.f., the CIMOSA cube has 36 combinations) therefore further simplification is highly desirable.

17.3 ELIMINATE UNWANTED COMBINATIONS

Every combination of the domains of the above dimensions defines a model category (description category) and each of the CIMOSA, GRAI-GIM, and Purdue etc., architectures cover a group of these.

The name of a description (model) category is rather long, although the name conveys a precise meaning, e.g.

> 'Generic function requirements of the human software (i.e. skills) pertaining to the CIM system mission (i.e. customer service)'.

Unfortunately too large a number of combinations are conceivable, thus the presentation could become too complex to be enlightening.

At the same time we can observe that a number of the categories have no practical meaning, or no practical meaning alone, so this number can be substantially reduced. We do this reduction through the use of the following four steps:

1. We notice that the generic architectures involved are each talking about two different kinds of 'product' each of which should manifest. Product one: a CIM project and product two: a CIM System.

   We should investigate separately what generic blueprints these architectures develop for each of the above.

   (a) A CIM Project Architecture

   (b) A CIM System Architecture

   For both the CIM Project and the CIM System there might be developed a separate form of the matrix for analyzing how much and to what degree each of the investigated architectures has covered the subject involved. E.g. the Purdue 'Implementation Procedures Manual' could be discussed on a Project Evolution matrix, while the 'Purdue CIM Reference Model' could be discussed on a System Architecture matrix.

   Thus there might be two matrices, one matrix for the 'CIM Project' and one matrix for the 'CIM System Architecture'.
2. The division into hardware and software is not meaningful at the requirements level except the resource aspect, this is because a functional specification abstracts away from the way the function is implemented; the above also applies to the human/machine division.

   The division into hardware and software is meaningful from the resource aspect as it is possible to separately define hardware resource requirements and generic software requirements.
3. The division of resources into the categories of computer, human and machine is adequately handled by the division into human or machine and into control and information or customer services.

An approach to proceed further is to define the n-dimensional space of model categories irrespective of their use in any of the architectural proposals, and mark those points with a '—' mark or cross in that space where the particular category is not used or not applicable.

4. Software and Hardware in the context of non-computer machines or humans can be a useful division following the principles developed in SATT. The elements which form the physical structure of a system can be divided into:
   (a) Entities of primary physical existence (hardware). Primary resources (entities) are material processing machines, computers, humans (used separately or together to carry out a function). These resources may have an independent physical existence.
   (b) Entities of secondary physical existence (software). Secondary physical entities are the identifiable, controllable parts of a hardware's state, typically brought about in order to implement some functions. These resources (entities) are called secondary, because they do not exist independently of primary entities. They are concrete resources nevertheless.

   Example (1): computer software is a state of computer hardware – e.g., a program, a database.

   Example (2): human software is either a state of a human (e.g., skills) or an association of humans (like an organizational entity, a department, a team...).

   Example (3): non-computer 'software' could be a particular machine configuration (e.g., a given cell configuration in a reconfigurable manufacturing system.)

17.4 DEVELOPMENT OF THE MATRIX FORM

The initial form of the proposed matrix is presented in Fig. 17.1. It was intended to be a representation of the taxonomy of the models used by the several architectures. This matrix form became the source of considerable discussion by the members of the Task Force for the following reasons:

- The reasons for the selection of the various dimensions of the matrix and how they would be used by the evaluators were not clear to the members.
- Was a taxonomy of models the real evaluation criteria which we wanted? Did it really tell us how valuable an architecture would be or how well it would help the user in designing, implementing and using an enterprise integration system? Since the 'life-cycle' was considered so important by the Task Force members that it was the major criteria for our initial

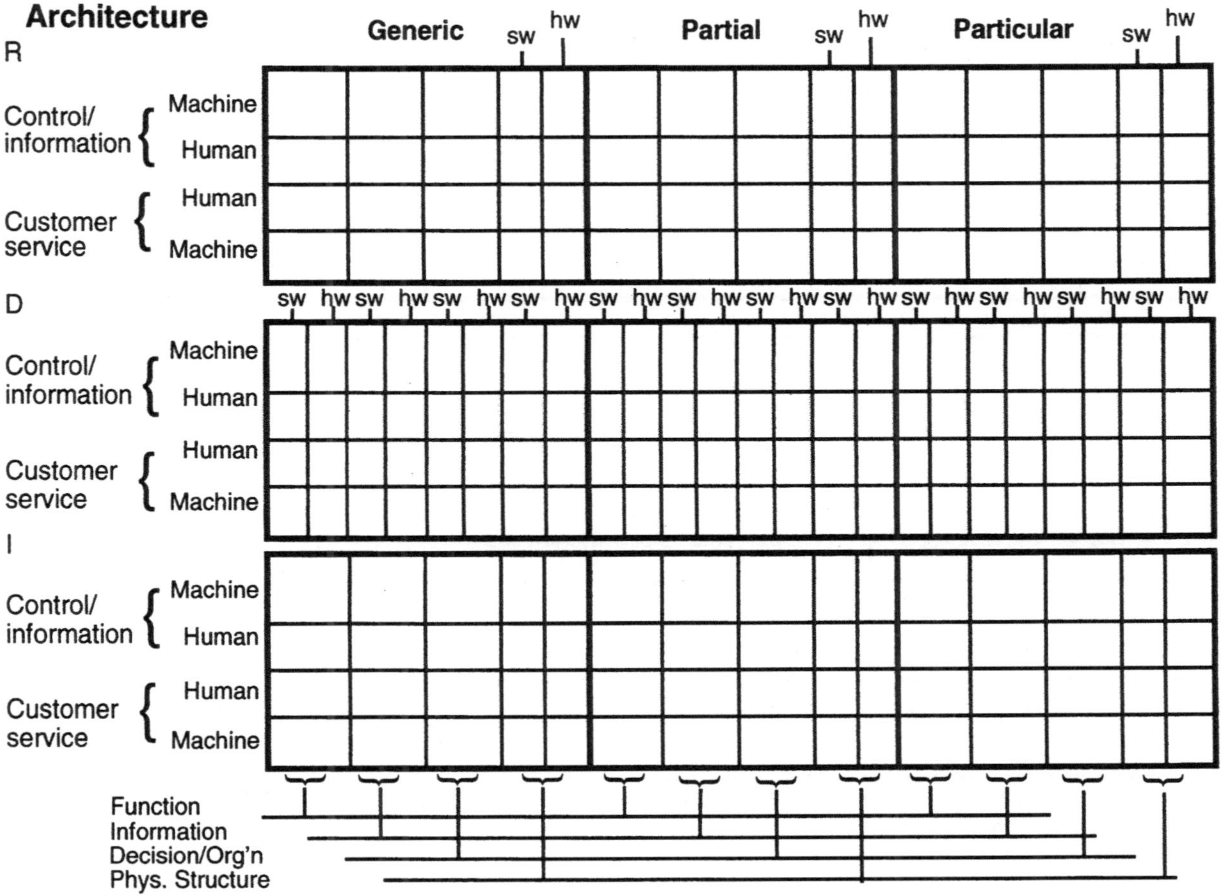

Figure 17.1 Initial proposed matrix for CIM system and methodology architecture evaluation.

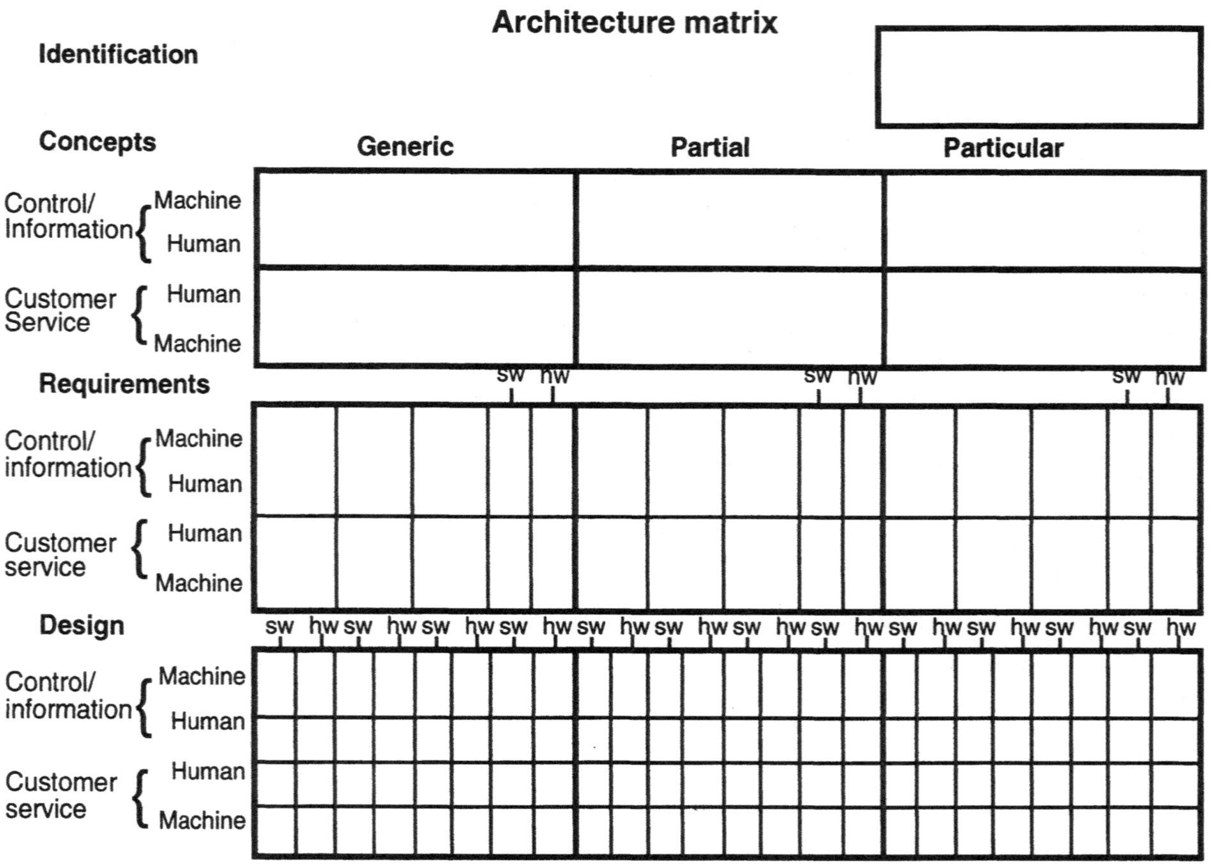

Figure 17.2 Final form of the evaluation matrix.

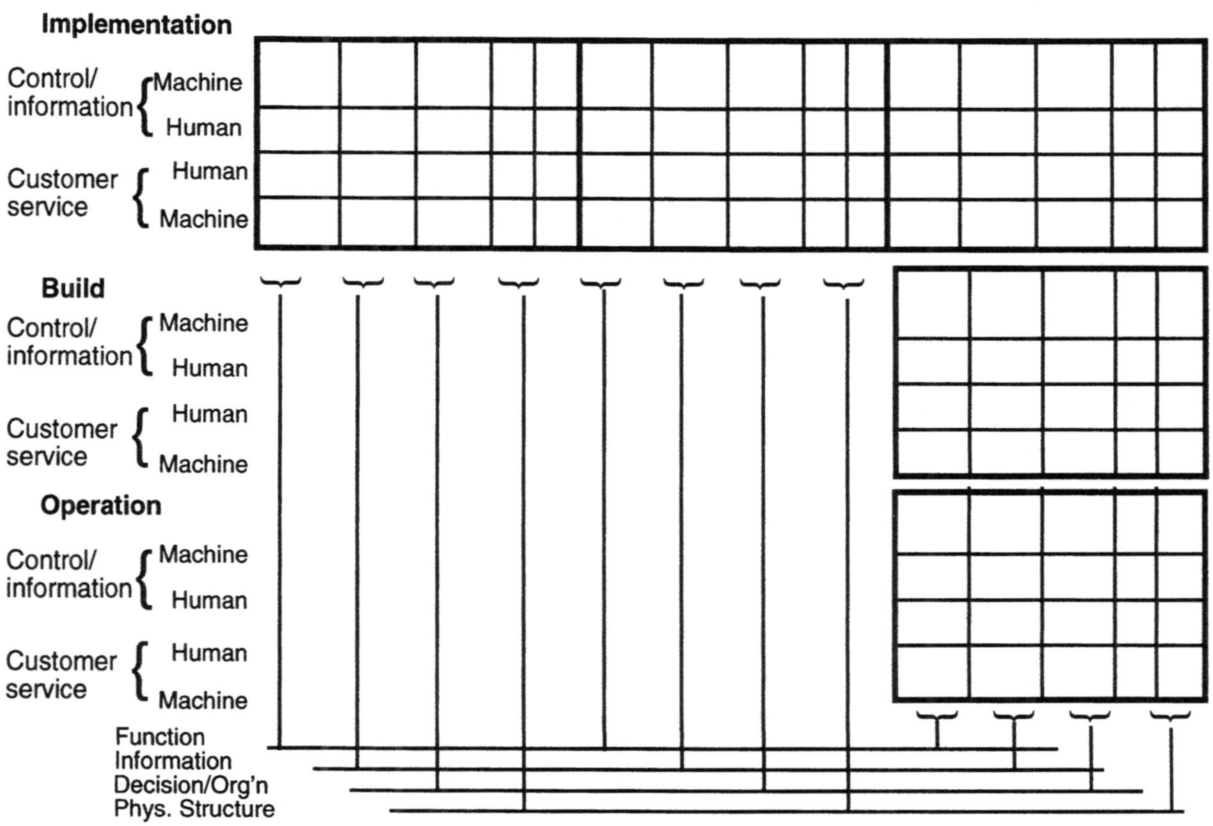

Figure 17.2 Final form of the evaluation matrix (Continued).

selection of architecture to be evaluated, should it not become a major part of the matrix structure? The following consideration also entered into the succeeding discussions which followed. Certainly integration of the development and use of information from both internal and external sources of the enterprise is important, but in addition, integration of the mission activities of the enterprise (e.g. physical manufacturing) is equally important, such as machine placement, material routing, etc.

There are also other considerations in the design of the matrix form which could simplify the presentation further.

In terms of project evolution or life-cycle, genericity can only apply to requirements and functional design or functional specification. Detailed design must always consider the specific requirement of a particular enterprise if it is going to be meaningful.

Certainly tools and techniques can be and are generic, but given a set of functional specifications these are the same whether one is designing a CIM system or any other equivalent enterprise undertaking.

Thus in terms of generic and partial modelling, the detailed design level of Purdue, and the equivalent implementation levels of GRAI-GIM and CIMOSA are meaningless.

If the above is true, then the manifestation and operations layers of Purdue and the corresponding build and operate layers of GRAI and CIMOSA are also particularly specific in each case.

Project evolution must apply to a specific CIM project. Again as above, tools and techniques would not be restricted to CIM projects alone but could apply to any enterprise project of whatever nature. In addition, achievement of the desired integration (both of information and mission components) is the distinguishing element here. Thus project evolution should be confined to the particular level of CIMOSA.

As a result of these and other discussions the matrix form was amended as shown in Fig. 17.2.

17.5 A WAY TO SITUATE A GENERIC ARCHITECTURE IN THE MATRIX

The answers given in the evaluation questionnaire (Chapters 11-15) help in correctly characterizing an architecture but, without consulting the authors of the architecture or reading in detail the definitive documents, we cannot hope to be sure of our interpretation of the answers.

Given this warning we filled in the Matrix as follows. We first identified which areas of activity are present and from what respect they are separately dealt with in the given architecture. Then we asked ourselves the following question (demonstrated on the Purdue Architecture as an example):

17.5.1 What descriptions will the direct or indirect users of this generic architecture have to produce when entertaining an integration project?

This was easy to answer because the requested descriptions are identified as deliverables on the Purdue two tier diagram. Note that the same list of deliverables could be identified from Question 17 of the Evaluation Questionnaire, Version 3.

17.5.2 What area of the matrix do these descriptions fit?

For each description (or sometimes a group of descriptions) above we circled an area in the matrix. (See Fig. 17.3 for a description of the symbology used.) For an example, let us consider the left hand side of the Functional Design or Specification Layer of the Purdue Architecture as given in Fig. 16.11 (same as Figure 8.8a and b). This shows the Information Systems Architecture (machines) and that part of the Human and Organizational Architecture (humans) which define the Information architecture at this level. We wish to determine how this area would be represented on the Evaluation Matrix of Fig. 17.2. For this purpose let us look at Fig. 17.6 which shows the representation of the whole Purdue Enterprise Reference Architecture on the Evaluation Matrix. This figure has also identified the areas numerically for the explanation (see Table 17.3) which will help our explanation here.

We therefore wish to discuss the area covered by ellipse 15 (we are discussing particular models). Based upon the vertical dimensions of the Matrix (as labelled at the bottom of Fig. 17.6) and the horizonal dimensions – upper (Control/Information) half of the R (Requirements) layer of the matrix, we can make the several observations concerning ellipse 15. This symbol occupies the following areas of the Matrix:

(a) Functional requirements of control/information functions to be carried out by humans.

(b) Functional requirements of control/information functions to be carried out by machines (computers and control equipment).

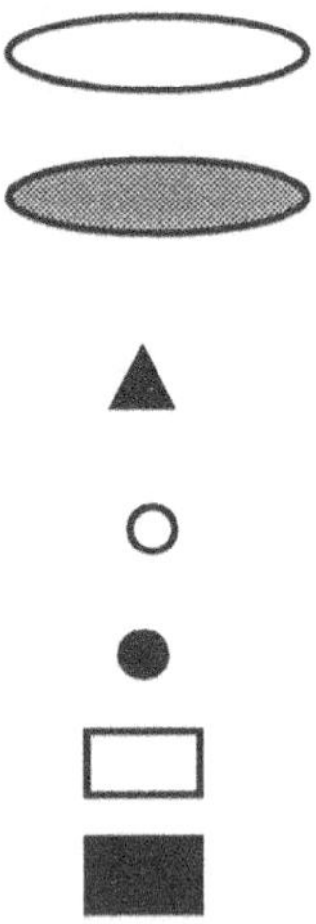

Separately defined description category; methods not defined

Separately defined description category; with extensive methods and design principles available

The architecture comprises a definitive model of this aspect

No formal modelling language is proposed but indicates some which would be adequate

Prescribes formalized modelling language to be used for all modelling activities

Identifies system component (module)

Detailed module specifications exist

An area surrounded by a solid line corresponds to a type of description which is separately defined in the architecture.

If the area is shaded, then the architecture also provides – in addition to defining the category of description – an extensive set of guidelines which can be used to facilitate and speed up the generation of the necessary descriptions (models).

A solid triangle within such an area means that the architecture incorporates a generic model which can be utilized as a blueprint by the user of the architecture, or could be the basis of standardization. An unfilled triangle means a model which is specified but not provided (yet) in detail in the architecture.

A circle within the area (marked with solid line or with shading) means that the architecture proposes or prescribes a modelling method which can be used as a means of representation for the given category. The circle is filled if the modelling language is formal and unfilled if it is only partially formal.

A rectangle within a marked area represents a module description (thus is meaningful only in the resource aspect). An unfilled rectangle means a software or hardware (or aggregate) module is present in the architecure definition as a separate identifiable component of the implementation. A solid rectangle means that a definitive detailed module description is part of the architecture (e.g. the Information Integration Infrastructure of CIMOSA).

Figure 17.3 Symbology used for completing the matrix form evaluations.

(c) Information requirements of control/information functions to be carried out by humans.

(d) Information requirements of control/information functions to be carried out by machines (computers and control equipment).

(e) Decisional/organizational requirements of the control/information functions to be carried out by humans.

(f) Decisional/organizational requirements of the control/information functions to be carried out by machines (computers and control equipment).

Note that the same list of deliverables could also be identified from Question 17 of Questionnaire, Version 3.

Ellipse 15 is shaded since Purdue does provide a 'separately defined description category with extensive methods and design principles available.' The symbol also contains a small open circle to indicate that Purdue: 'does not propose formal modelling language(s) but does indicate some which would be adequate.'

17.5.3 Does the methodological handbook (or other manifestation of the methodology) tell the user in detail how to produce these descriptions?

We tried to answer this question for each area in the matrix.

If substantial and detailed methods were in evidence i.e. the methodological handbook told the user how to produce the given description, then we shaded the area.

17.5.4 What kind of descriptions of the enterprise or any component thereof is delivered with the architecture documents?

Obviously it is of great help for someone who starts an integration project if there are standard, paradigmatic, prototypical, or blueprint solutions which can be taken over and applied as they are or in an adapted form.

The above can be delivered in two forms:

(a) An abstract 'generic' model (like the ISO OSI model) which can be filled in with concrete detail according to the individual case.

(b) A concrete prototypical solution, of which parts are standardized and where the non-standardized parts can be changed according to the individual case.

There is evidence that in fact the latter is more usable for any (human or automated) design activity.

We identified those generic (abstract or concrete) models which came ready with the architectures. A typical example for this is the Purdue Reference Model for manufacturing control. Each of these (not very numerous) models are represented in the Matrix as a triangle.

Again, obviously, we want generic architectures with lots of triangles in them, because the user will then have many example cases to build on.

When mapping architectures (e.g. Purdue [P] and CIMOSA [C] we need to worry about two questions:

- Is there a triangle for the same description category in both P and C? If yes are they compatible, overlapping, equivalent — interestingly and relevantly we have found no such problem as yet.
- Can what a triangle of P represents be made useful in C? The answer is definitely yes – provided the form of expression of P can be semantically mapped to the form of expression of C. Thus for each of the triangles it is possible to define a project which would translate (using whatever efficient means can be found) the triangles in P to C or vice versa.

17.5.5 What formal ways of modelling are prescribed or advised by the architecture?

For each description category we asked (again using Question 17 of the Questionnaire, Version 3): Is there a modelling language defined (prescribed) at all and if yes is it a formal modelling language (with a mathematically defined semantics) or is it a semiformal one (with some formalism but no mathematically defined semantics)?

If semiformal then an empty circle was drawn in the area, if formal then a filled circle. On occasion the circle was drawn next to a model triangle signifying that the model is indeed provided in the intended formalism.

17.5.6 Does the architecture intend to build on a generally available hardware/software module/infrastructure?

(This information is partially available from Question 12 of the Questionnaire, Version 3.)

In the case of the Purdue architecture no such common module is defined, however recommendations are made for the use of such common modules as OSI.

In case of CIMOSA the Information Integration Infrastructure is defined as such a common module.

Where alternative modules could be used as opposed to one standard prescription it may be useful to group them together as a set of possible solutions but we have not made provision at the moment for graphically presenting this in the already complicated matrix.

Again, there are two possibilities to mark the modules defined by the architecture.

(a) If the module is defined as a necessary component but the architecture does not provide detailed specification of the module then this fact is represented by drawing an empty box in a respective 'resource' column of the matrix.

(b) If a module is not only defined to exist but also specified (such that it could be standardized) then we draw a filled area box into the respective 'resource' column of the matrix.

17.5.7 How does the architecture describe how an integration project should be presented (what it should look like)?

We notice that this question is not about the integrated enterprise itself. It is about a project to create one. Just as the enterprise has a structure, a project also has one. This structure, if looked at from different viewpoints, forms an architecture. We call this the 'Architecture of the Methodology' or the "Architecture of Project Evolution"; therefore we asked:

(a) Is there a model of integration projects available (▲)?

The CIMOSA Life-cycle model or the Purdue Implementation Procedures Manual are such generic descriptions which pertain to the service that an integration project carries out. It is to be investigated how the control of integration projects is to be specified. Are there any special types of requirements or any generic project management methodology which is partiocularly adequate?

(b) Are the descriptions of the project identified (⬭) ?

It is not clear at the moment if any of the architectures require that the project be modelled in a given way (●) or that methods would be made available for generating such descriptions (⬬)

17.5.8 What are the methodologies used to carry out integration projects?

The existence of methodologies is shown on the matrices by shading the areas of descriptions required in the respective architecture.

17.6 A SET OF MATRIX FORM EVALUATIONS

Fig. 17.4 presents the mappings of CIMOSA on the matrix. Accompanying the diagram is an explanation of the mapping of each area in Table 17.1. Fig.17.5 presents the matrix evaluation form for GRAI-GIM, Table 17.2 shows the corresponding explanation. Please note the reference back to the GRAI Laboratory's explanations of the one-on-one mappings of Chapter 16; these are considered as additional explanations of the similar decisions made in the matrix evaluations. Similarly, Fig. 17.6 presents the corresponding information on the Purdue Enterprise Reference Architecture and the Purdue Methodology along with Table 17.3.

Table 17.1 Definition of the areas in the CIMOSA matrix evaluation form of Fig.17.4

#	*Explanation*
1	Generic functional requirements model (functional requirements of IIS – ▲ ; CIMOSA modelling construct – ●)
2,3,4	Generic information, organization, resource requirements models
5–8	Same as 1–2 except partial
9–12	Same as 1–2 except particular box in 12 represents IIS
13–15	Same as 1–4 on the design level Shaded area represents that methods for software design only are available A shaded area not included in Item 16 since no method for the design of a resource structure are available
17–24	Correspond to the discussions concerning Items 13–15 except for partial and particular levels

Table 17.1 Definition of the areas in the CIMOSA matrix evaluation form of Fig.17.4 (Continued)

#	*Explanation*
25	Detailed design of application programs (the shaded area)
13–32	Presentation shown is limited to a discussion of the enterprise software required in view of the fact that the CIMOSA scope may be limited only to this area
	(a) CIMOSA modelling languages are discrete event methods, therefore targeted to discrete manufacturing; the range of detail necessary for continuous processes is outside the scope of CIMOSA
	(b) With the assumption of embedded control (as in NC machines, etc), the scope of CIMOSA has only a restricted applicability in industries other than discrete manufacturing, which do not make these assumptions
	(c) Software design methods only are available in CIMOSA (hardware and manufacturing equipment design methods are outside the scope)
35	As earlier concluded, the existence of generic requirements for the physical part (equipment) of a factory is unanswerable since no information is available to identify factory or its nature and behavior (there is no such thing as 'general factory'); however, generic constructs (●) may be defined for modelling purposes
36	Likewise for generic design
37	Likewise for detailed design – the identity of the system involved must be known, ie be nongeneric for a detailed design to take place

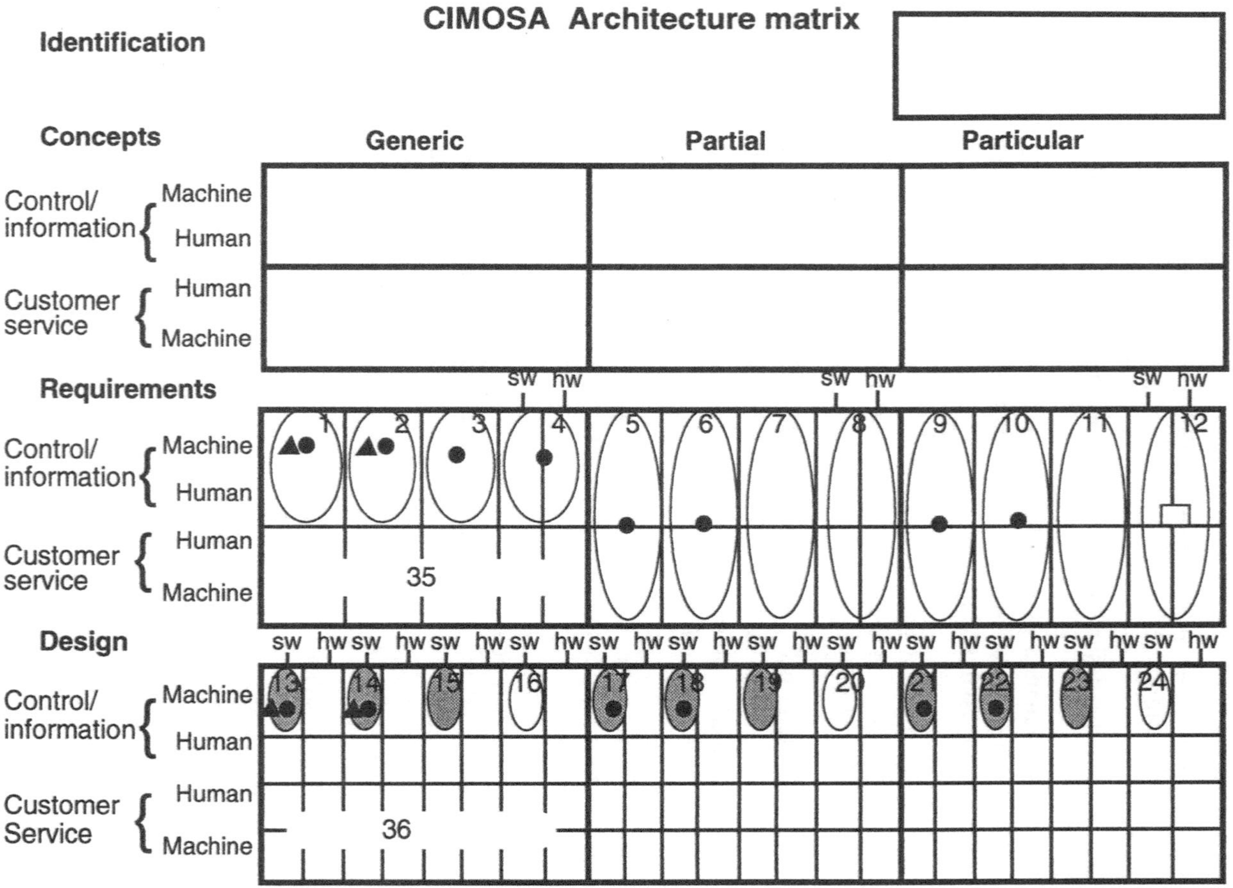

Figure 17.4 The matrix evaluation form for CIMOSA (see explanation of numbering in Table 17.1).

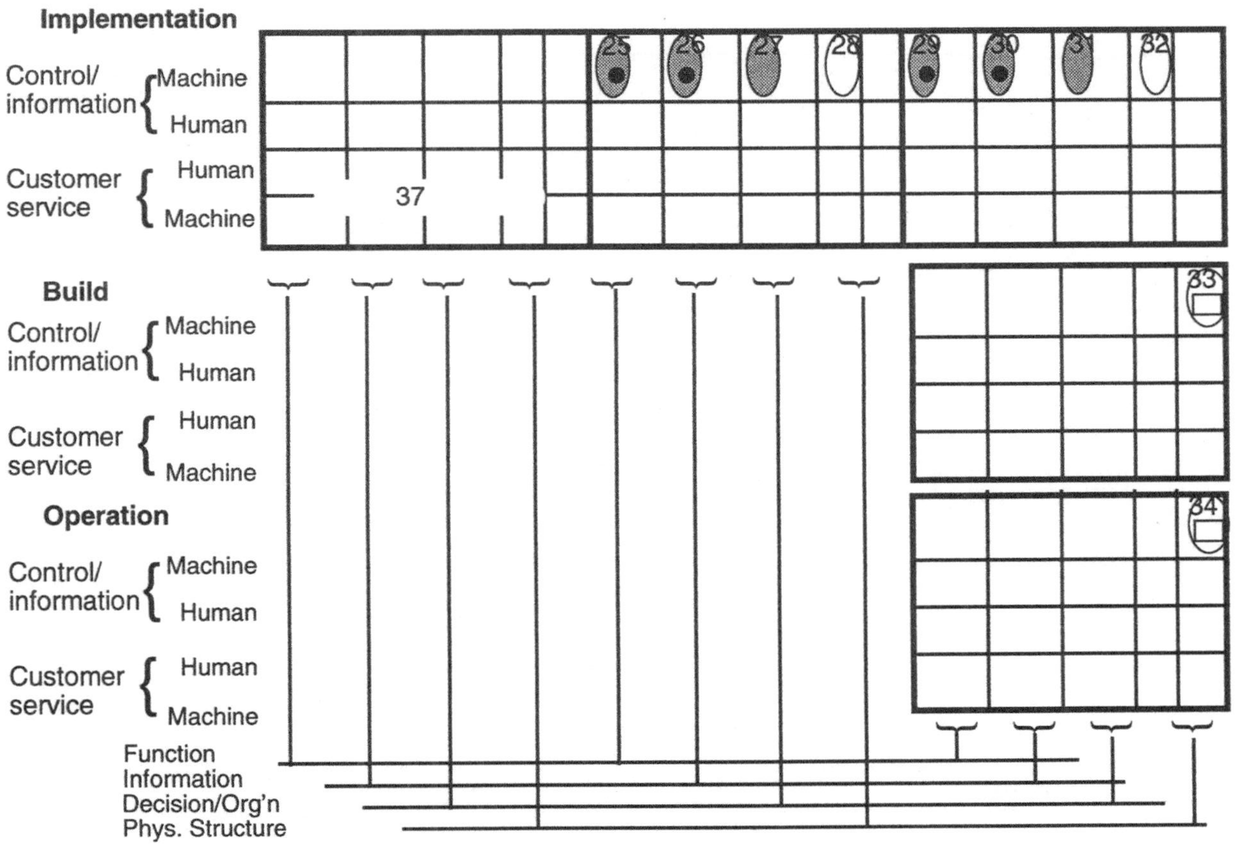

Figure 17.4 The matrix evaluation form for CIMOSA (see explanation of numbering in Table 17.1) (Continued).

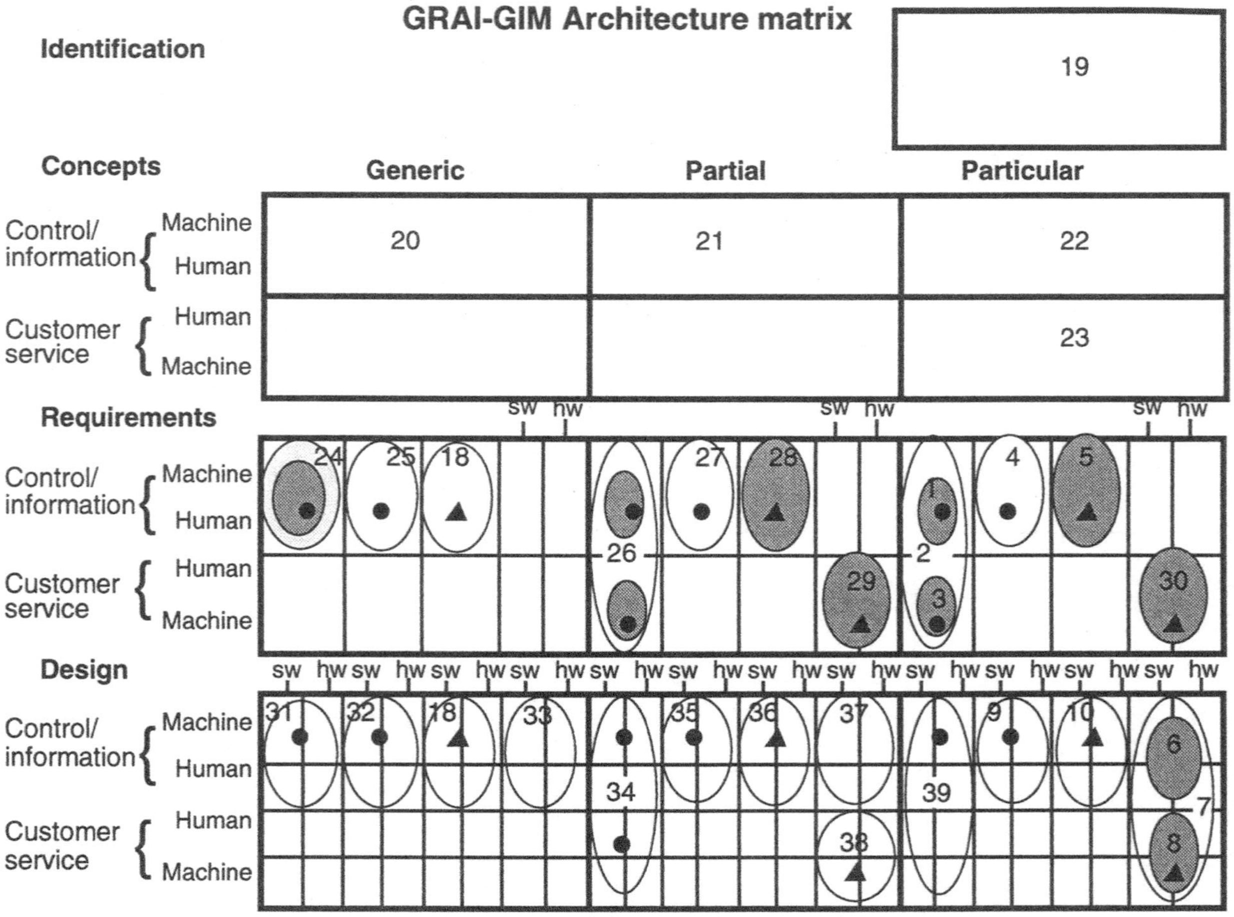

Figure 17.5 The matrix evaluation form for GRAI-GIM (see explanation of numbering in Table 17.2).

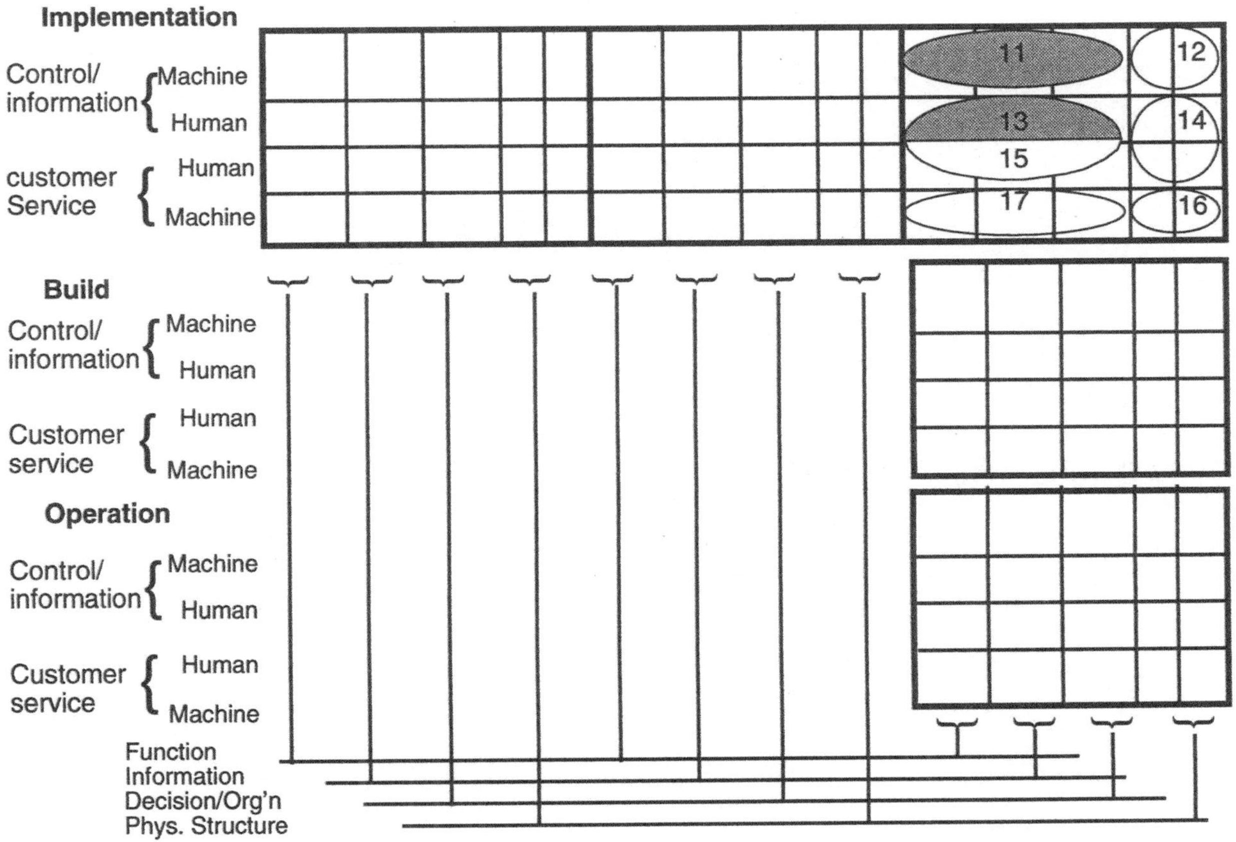

Figure 17.5 The matrix evaluation form for GRAI-GIM (see explanation of numbering in Table 17.2) (Continued).

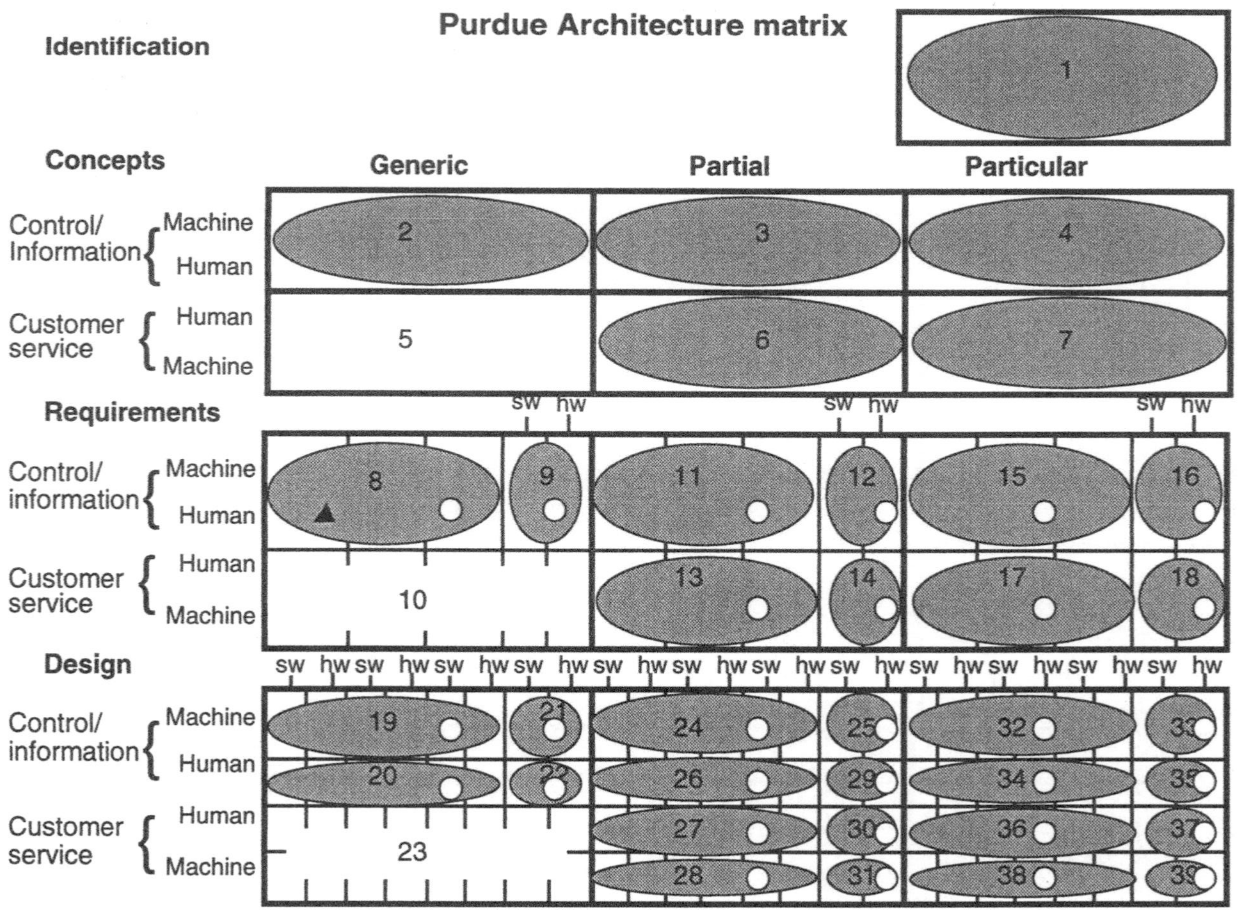

Figure 17.6 The matrix evaluation form for PERA (see explanation of numbering in Table17.3).

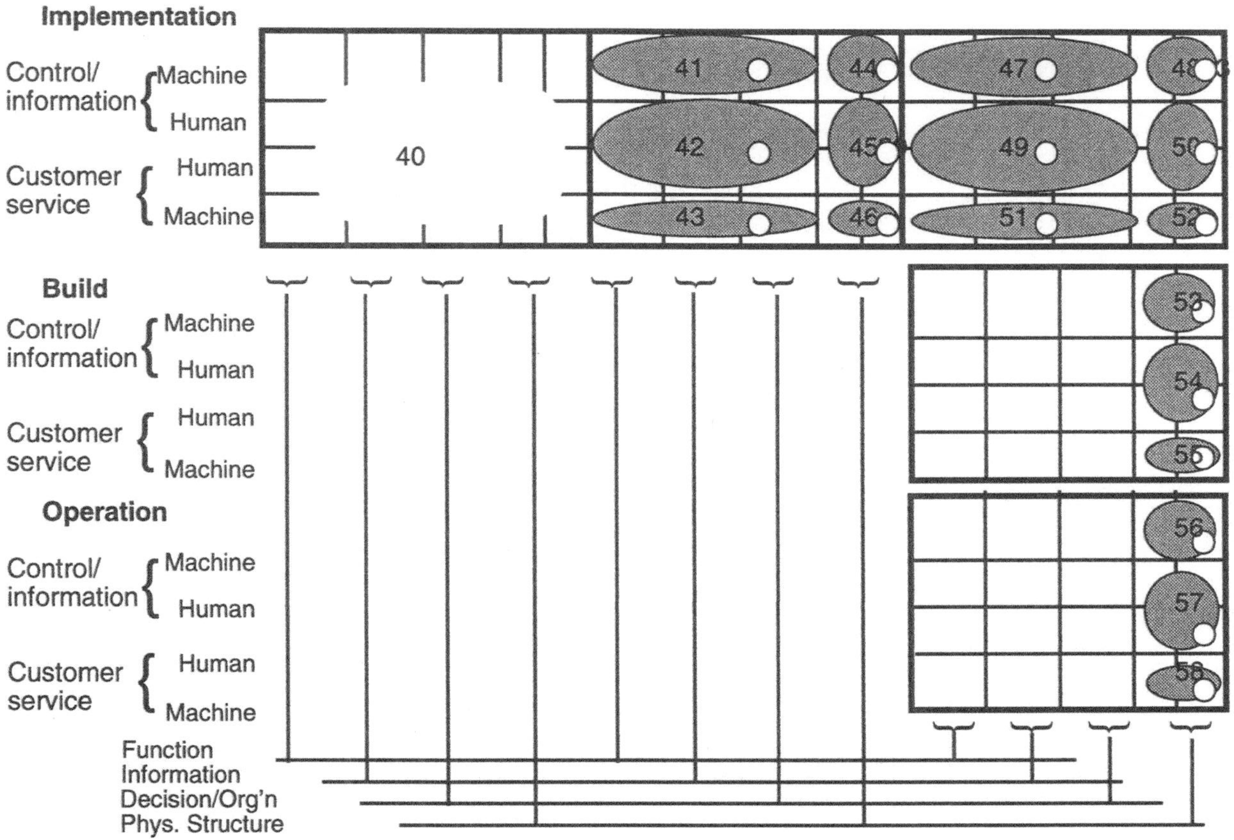

Figure 17.6 The matrix evaluation form for PERA (see explanation of numbering in Table 17.3) (Continued).

Table 17.2 Definition of the areas on the GRAI-GIM matrix evaluation form of Fig.17.5

#	*Explanation*
1	Functional view of control system as subsumed by the decisional view. Item 1 is part of the decisional model which describes the control functions (1 and 5 are parts of a single document described in the formalism of GRAI-grid and GRAI-net)
2	Functional view – IDEF0 diagram of the entire enterprise
3	Physical view – IDEF0 of the manufacturing (physical) processes including physical facilities and products [GIM conceptual physical model – CPm]
4	Information view – entity relationship model [GIM conceptual information model – CIM] (GRAI-GIM #4)[a]
5	Items 6–10, same as above but now for Detailed Design (Structure) Instead of Functional Design or, Specification (Function)
6	is subsumed by Item 10 (ie, GRAI-GIM #6 is a model which contains the structural design of the control system)
7	Item 2 with mechanisms indicated [GIM structural functional model – SFm]
8	Item 3 with resources and structure indicated [GIM structural physical model – SPm] (GRAI-GIM #5)
9	[GIM structural data model – SIm]
10	[GIM structural decisional model – SDm] (GRAI–GIM #6)
11,12	Detailed design [GIM implementable information technology model – IITm]

Table 17.2 Definition of the areas on the GRAI-GIM matrix evaluation form of Fig.17.5 (Continued)

#	*Explanation*
Part of 13,15 and 17	[GIM implementable organizational model – IOm] (GIM #7)
Part of 14,16	[GIM implementable manufacturing model – IMTm] (GIM #8)
18	Generic production management model
19	Announcement of business entity involved
20–23	Definition of the domain of the study
24	Function view partly described by IDEF0 and partly described by GRAI grid and GRAI nets
25	Entity relationship modelling formalism
26	Partial function view model partly covered by IDEF0 and partly described by GRAI grid and nets
27	Partial information model using the formalism defined in Item 25
28	Partial decision model using the formalism defined in Item 18
29	Partial S/R (stock/resource) formalism to model physical subsystem transformation process
30	Particular S/R model

Table 17.2 Definition of the areas on the GRAI-GIM matrix evaluation form of Fig.17.5 (Continued)

#	*Explanation*
31–39	Same as defined in the requirements level but distributed detailed design; at the technical design stage, the four views, namely function, information, decision and physical are transformed into three categories of components: information technology components (hardwares and softwares, databases, LAN etc), manufacturing technology components (machines and human etc) and organization components (not tangible such as organization structure, management rules, working procedure etc)
33 and 37	Same as 6

a. These parenthetical expressions [ ] and (GRAI-GIM #) refer to Fig. 16.21 and 16.22 and to the numbered superscripts used in Fig. 16.15.

Table 17.3 Definition of the areas in the Purdue Reference Architecture's matrix evaluation form of Fig. 17.6

#	*Explanation*
1	Identification of the CIM business entity involved as called for in the Purdue Enterprise Reference Architecture
2	Mission, vision, and values; goals and objectives; etc, of management for the information part of the contemplated enterprise integration project; there are many such goals, etc, which are generic to all situations
3	Once the industry involved has been specified there are additional goals and objectives, etc, which become common; here these are for the information system part of the overall enterprise

Table 17.3 Definition of the areas in the Purdue Reference Architecture's matrix evaluation form of Fig. 17.6 (Continued)

#	*Explanation*
4	Once the specific CIM business entity is specified, the complete mission, vision, values, goals and objectives, etc, can be specified; as above these are for the information systems part of the enterprise
5	Without either a specific plant or industry having been specified, generic goals, etc, for the manufacturing part (customer service) would not seem to be definable
6	Similar to Item 3 but for the customer service part of the enterprise
7	Similar to Item 4 but for the customer service part of the enterprise
8	The functional requirement part of the information system; this has been shown repeatedly to have many generic aspects regardless of industry involved
9	Structural aspect of the functional requirements; again there are many proposed generic solutions for the information system itself regardless of application
10	For the customer service part the genericity aspects of the solution are not present
11	As with Item 3 above, designation of the industry involved make a much larger selection of common practices available for exploitation by the user
12	Designation of the industry involved makes additional common features for the structure of the information system available to the integration planning team
13	Similarly to Item 6 above, establishment of the industry involved permits much commonality of the manufacturing equipment to be exploited

Table 17.3 Definition of the areas in the Purdue Reference Architecture's matrix evaluation form of Fig. 17.6 (Continued)

#	*Explanation*
14	Like Item 12, designation of the industry involved will uncover commonalities in the structures of the customer service (manufacturing) elements of the Enterprise as well
15	Designation of the business entity involved makes the full details of plant, project and information system available for requirements analysis and design specifications
16	These are functional requirements for the structure of the information systems; to be valid at this point they must have been mandated by industry or company commonality, management mandate, or mandates coming from national or local laws and other such requirements
17	Designation of the business entity involved plus full details of plant production goals and objectives make the preparation of complete plant design requirements attainable
18	Just as with the information systems functional requirements, mandated requirements from governmental laws or rulings, standards adopted or management instructions will dictate requirements for plant structure and organization even at the functional requirements level
19	The genericity of control systems today is such that considerable functional design of information systems elements can take place even at the generic level – such as components of a distributed control system or computer elements themselves for the control system
20	With the human versus machine split taking place at this level, the functional requirements for the human assigned portions of the generic control would be attainable at this level
21	Genericity capabilities for control systems can also permit the separate elements of the systems to be part together in generic structure even with the industry or specific plant unknown

Table 17.3 Definition of the areas in the Purdue Reference Architecture's matrix evaluation form of Fig. 17.6 (Continued)

#	*Explanation*
22	Similar results to those for the hardware and structural elements of control systems are also possible; here is direct relationship to the statements of Items 19, 20, and 21
23	While considerable genericities are available in control system design, functional design of the manufacturing system does require a considerable knowledge of the plant to be available before such work is potentially useful, thus this requirement would be undefined at this point
24–31	Directly analogous to Items 11–14 for functional requirements and to Items 19–22 for the information systems elements of the generic block
32–39	Again directly analogous to those just completed at Items 24–31 and also with Items 15–18 above
40	Detailed design requires a very complete knowledge of the planned enterprise; therefore detailed design would not be definable for the generic level, except for tools which might be available generically for control system design at more specialized layers
41–52	The work at the detailed design level will again follow the pattern already established above; industry knowledge will allow some detailed design even at the partial level, particularly for control systems but also, to a lesser degree for production systems as well; note that human applications design functions are considered as a whole at this level since concentration here is on skills and training, etc, as a whole
53–58	The build and operate levels are strictly a particular system function since one only builds and operates a specific plant – all have their own individual characteristics which must be considered

18

General observations in comparing candidate architectures

T.J. Williams

18.1 SOME COMMENTS CONCERNING THE MODELLING FRAMEWORK USED BY CIMOSA

Questions arose several times during the meeting of the Task Force concerning whether or not all of the categories of models as depicted in the CIMOSA Modelling Framework (the cube) were populated, ie, was every possible model defined? Figures 18.1–5 present one possible interpretation of this question based upon the results of the Mapping Against a Matrix of Requirements of Chapter 17.

18.2 SOME COMMENTS ON THE IMPACS MODELLING FRAMEWORK AS USED BY GRAI-GIM

Although the definitive description document on the GRAI-GIM Methodology [1] does not specifically describe the origin of this methodology to be in the IMPACS Program of ESPRIT as carried out at the GRAI Laboratory, reading of their earlier publications certainly describes such an origin [2–4]. Thus it is helpful to review Fig. 18.6 from IMPACS in understanding Fig.18.8 as used in the GRAI-GIM Methodology (See Figures 7.13–15).

If one takes Fig. 18.6 and removes those blocks which are apparently undefined in the framework, one obtains the representation of Fig. 18.7. Since the IMPACS Figure is strictly a Particular Architecture (there is no mention of Generic and Partial constructs) it would certainly be less populated than the corresponding CIMOSA Framework diagram.

Important here is the fact that a representation of the Fig. 18.7 result as viewed in the direction of the observation arrow shown there gives us the two dimensional view of Fig. 18.8, which is extensively used in the GRAI-GIM Methodology as described in Chapter 7. This is helpful in understanding the latter figure and its use in Chapter 7 as one of the major bases for the GRAI-GIM Methodology.

18.3 GENERIC, PARTIAL AND PARTICULAR BUILDING BLOCKS AS SHOWN BY THE PURDUE ARCHITECTURE

A major innovation of CIMOSA, in relation to the other two candidate architectures described in this report, is its espousal of the gain to be obtained from a collection of those constructs and principles which can be considered as being generic to all factories. This is particularly so when one considers the scheduling, control and monitoring system involved. This factor is also the basis for the Purdue Reference Model for CIM [5] which also presents a major list of generic functions of this type common to not only all manufacturing plants but to all industries as well.

This commonality of functions is further enhanced if one restricts the Manufacturing plants to one particular manufacturing industry. The resulting enhanced list of functions and models become 'partial' models and expand on the 'generic' models mentioned above. The combination of the two (generic plus partial) then become the 'reference architecture' for CIMOSA and an initial aid in developing the 'particular' architecture for any specific plant as shown by the CIMOSA Modelling Framework.

Both GRAI-GIM and Purdue were developed strictly for the study of specific plants, projects and installation and thus can be considered only as 'particular' architectures in the CIMOSA sense. However, their basic structures (Figures 7.19 and 8.1) can be replicated like the Building Block layers of the CIMOSA Modelling Framework (Figures 6.5 and 6.6) as shown in Fig. 18.9 and used to show the degree of commonality possible with the Generic and Partial layers of the CIMOSA Modelling Framework.

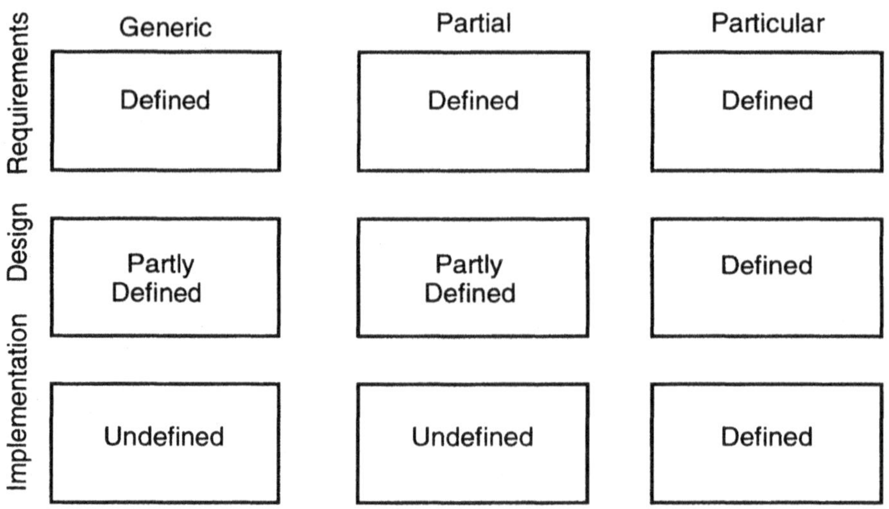

Figure 18.1
Population of the boxes of the information and function views of CIMOSA.

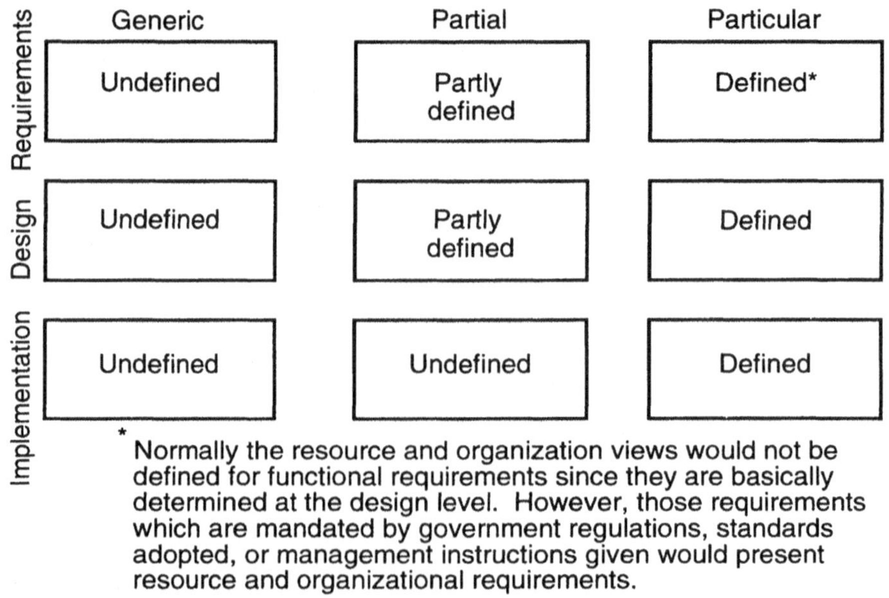

Figure 18.2
Population of the boxes of the resource and organization views of CIMOSA.

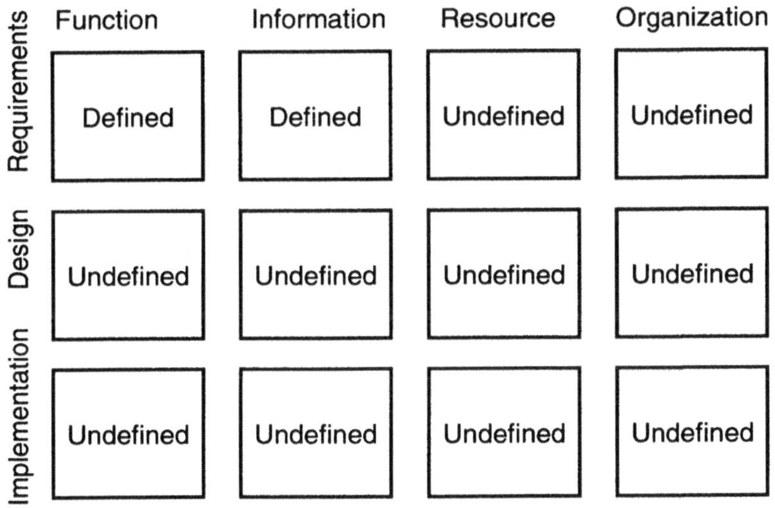

Figure 18.3 Population of the boxes of the generic level of CIMOSA.

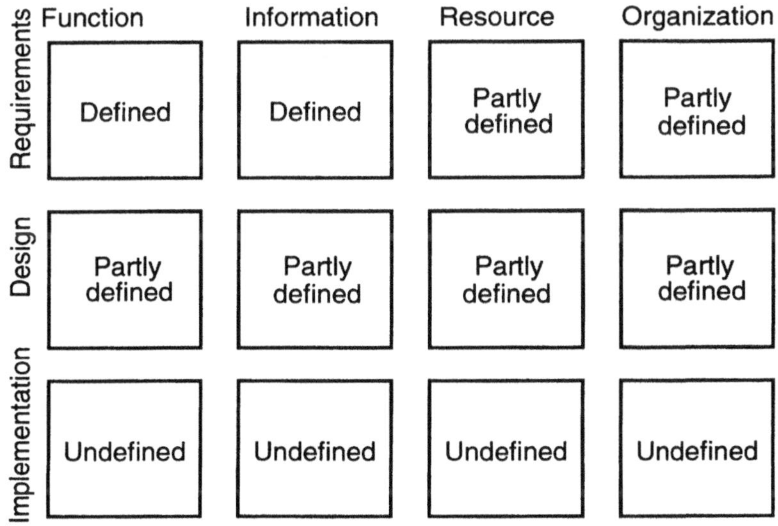

Figure 18.4 Population of the partial level of CIMOSA.

	Function	Information	Resource	Organization
Requirements	Defined	Defined	Defined	Defined*
Design	Defined	Defined	Defined	Defined
Implementation	Defined	Defined	Defined	Defined

Figure 18.5 Population of the boxes of the particular level of CIMOSA (* see note discussing the resource and organization views, Fig.18.2).

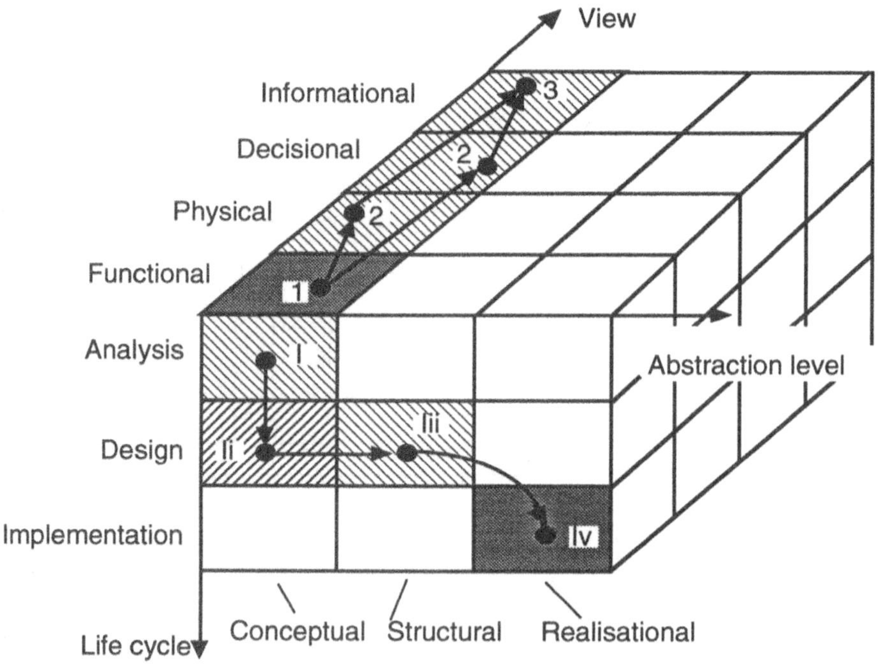

Figure 18.6 The IMPACS modelling framework of GRAI-GIM.

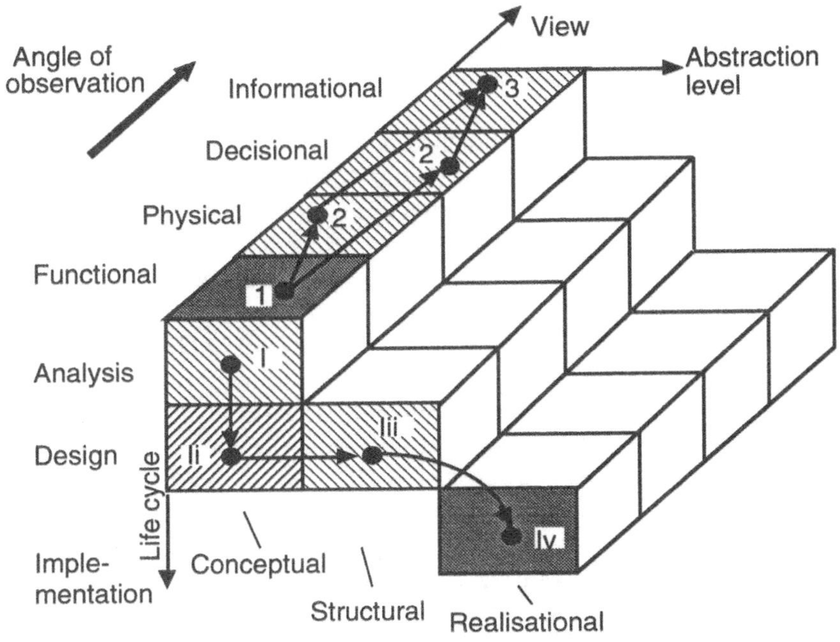

Figure 18.7 The development of the GIM modelling framework from the IMPACS modelling framework of GRAI-GIM.

Fig. 18.9 also presents a legend to be used in interpreting the information presented in this figure. The following notes are also important in its interpretation:

1. There is no 'Identification of the CIM Business Entity' in the 'Generic to All Enterprises (Generic)' and the 'Generic to Specific Industry (Partial)' Regions for obvious reasons since of course none would have been named in these types of studies.
2. There are no Detailed Design, Manifestation or Operations Layers in the 'Generic to All Enterprises' and the 'Generic to Specific Industries' Regions for the same reason as above.
3. In the 'Generic to All Enterprises' Region the Concept Layer is indicated to include the fact that many good management practices are common to all industries. The slanted line indicates these are information (control) functions. The same applies to the 'Generic to Specification Industry' region as well.

18.4 REFERENCES

[1] Doumeingts, G., et al, GIM-GRAI Integrated Methodology, Version 1.0 LAP/GRAI Laboratory, University of Bordeaux 1, Bordeaux, France (May 1992).
[2] Chen, D., Vallespir, B., and Doumeingts, G., 'GRAI Integrated CIM Architecture – A proposal'. Proceedings CIMCON '90: International Conference of CIM Architecture, Gaithersburg, Maryland, pp 153–165 (May 1990).
[3] Chen, D., Vallespir, B., and Zanettin,M., 'Architecture Globale CIM: Application dans le Project ESPRIT IMPACS' Proceedings CIM '90: Productique et Integraion, Bourdeaux, France pp. 243–250 (June 12–14, 1990).
[4] Chen, D., Vallespir, B., and Doumeingts, G., and Zanettin M. – 'The IMPACS Conceptual Reference Architecture'. Draft Version (November 1990). Report to ESPRIT, GRAI Laboratory, University of Bordeaux, Bordeaux, France (November 1990).
[5] Williams, T.J., Editor, A Reference Model for Computer Integrated Manufacturing (CIM), A Description From the Viewpoint of Industrial Automation, Minutes, CIM Reference Model Committee, International Purdue Workshop on Industrial Computer Systems, Purdue University, West Lafayette, IN (1988), Instrument Society of America, Research Triangle Park, NC (1989).

Abstraction level	Views			
	information	decision	physical	functional
conceptual				
structural				
implementable				

Figure 18.8 The GRAI-GIM modelling framework.

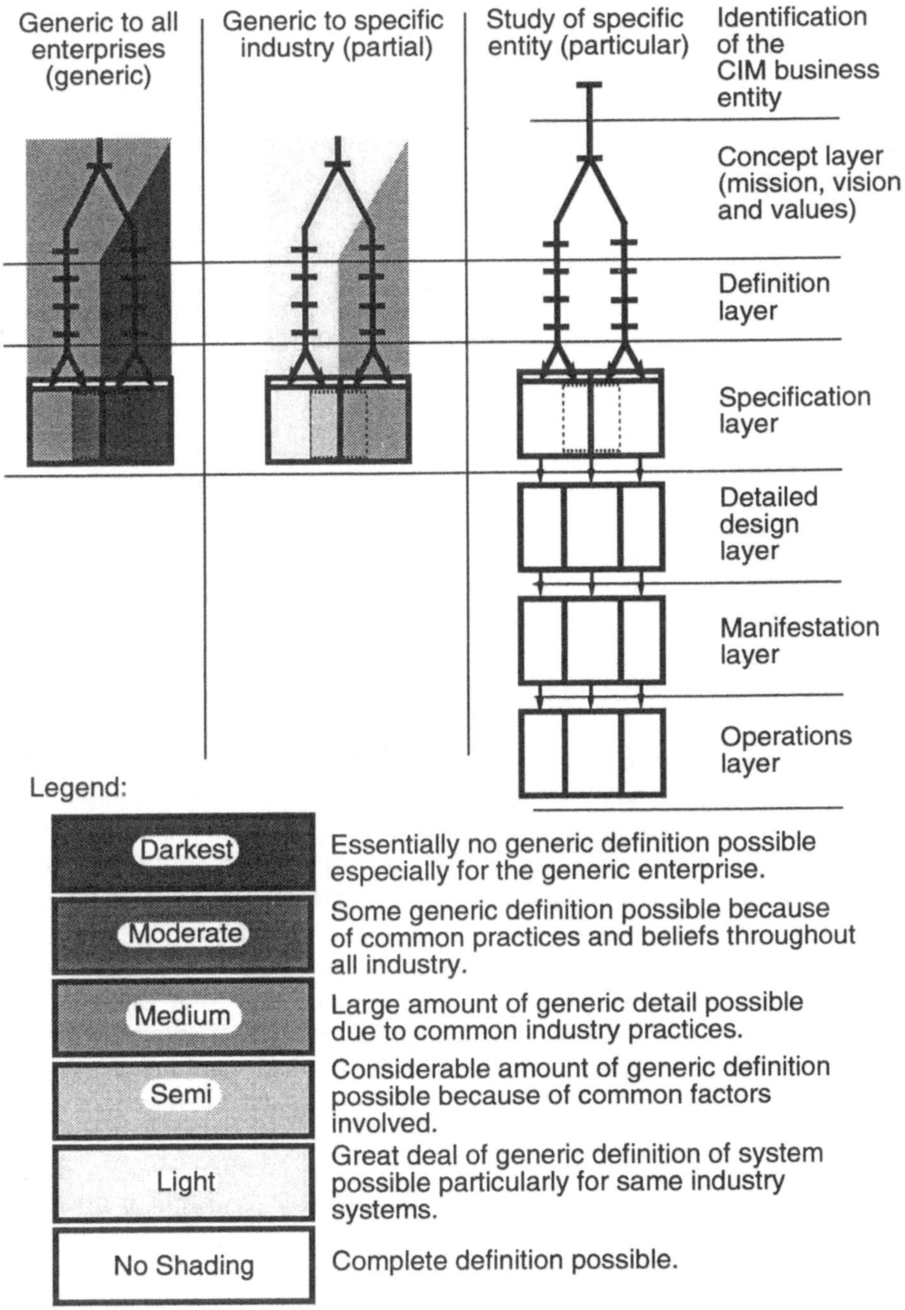

Figure 18.9 How the Purdue methodology handles the concepts of generic, partial and particular models as espoused by CIMOSA.

19

Conclusions and summary of findings

Task Force

19.1 GENERAL CONCLUSIONS

The following general conclusions are readily drawn from the work of the IFAC/IFIP Task Force on Architectures for Integrating Manufacturing Activities and Enterprises during the triennium 1990–93. Note that many of these Conclusions are also shared with those of several other groups concerned with companion studies to our own. Some of these have predated those done by the Task Force itself and thus are not necessarily new to the Task Force.

- There are a multitude of proposed architectures and models in both the open and proprietary literature which purport to illustrate, explain and guide the task of integrating manufacturing activities and enterprises. Thirty-five listed in Chapter 5 here alone. However, only a very few of these actually treat the 'how' of enterprise integration as well as the 'what' that is needed. The remainder concentrate on describing the structure of the computer control system involved itself and/or the interconnectivity of the various functions being carried out by it, i.e. the 'what' only.
- Thus, as just noted above, there are two basic types of enterprise integration architectures today: (l) those which describe the architecture or physical structure of some component or part of the integrated system such as the computer system or the communications systems and, (2) those which present an architecture or structure of the project which develops the integration i.e. those that illustrate the life history or life-cycle of the project developing the integrated enterprise.
- The experience of many CIM program groups worldwide has shown the value of a reference architecture in guiding the separate stages of planning, establishment, analysis, specification, detailed design, construction

and commissioning, and operation of projects for the integration of information and manufacturing activities in industrial plants and indeed all types of enterprises everywhere.

- Thus the most valuable architectures are those which model or describe the structure of the integration programs themselves, i.e. the developmental and application stages in the life history of the program. We shall label these as Type 2 architectures.
- These Type 2 architectures will include as vital component parts many examples of the other major type of reference architectures. These latter architectures, to be known as Type 1, model or describe the structure of the computer system and/or of other major components which comprise the overall integration system as finally developed. The vast majority of the so-called reference architectures in the literature today are of this second type.
- The Task Force determined that there were three separate architectures of the type described as Type 2 above which were important enough to merit detailed study by the Task Force. The Task Force was also fortunate in having members from the developmental groups for each of these three candidates architectures as members of the Task Force. This allowed our deliberations to profit from knowledge concerning these architectures not already in the open literature.

These three architectures were:

(a) CIMOSA – As developed by the AMICE Consortium under the ESPRIT Program of the European Community. ESPRIT Projects 688, 2422 and 5288, all entitled Specification of an Open CIM Architecture were involved (called CIMOSA below)

(b) GRAI-GIM – As developed by the GRAI Laboratory of the University of Bordeaux, Bordeaux, France, under its own research program for production management systems and with the aid of ESPRIT Project 2338 IMPACS (called GRAI-GIM below).

(c) Purdue Enterprise Reference Architecture and the associated Purdue Methodology – As developed by the Purdue Laboratory for Applied Industrial Control of Purdue University, West Lafayette, Indiana, U.S.A., under the work of their Industry–University Consortium for CIM (called Purdue below).

- The Task Force also determined that it is vital that the Type 2 reference architecture be accompanied by a methodology which gives detailed instructions to user project development groups on how to use the architecture to guide the conduct and progress of their study. The methodology should also detail the nature and use of all available techniques and tools valuable to the user group at each stage of the development and operation of the integration program and/or project.

- Enterprise integration is a very complex and highly detailed endeavor. Similarly, the reference architectures which describe this endeavor are also complex and highly detailed. As a result none of the candidate architectures and associated methodologies are, as yet, completely developed, described and documented. However, in each case the path to their ultimate completion is clear either from the work of the architectural development group themselves and their future plans or from the associated work of other groups, including this Task Force.
- Each of the studied architectures and their associated methodologies should be extended by the original development groups or others to become a complete architecture and methodology for guiding enterprise integration programs from initial conception to their actual construction and use. In our Recommendations section here the Task Force has outlined one method by which this could be accomplished in each case.
- It would also be possible, and perhaps more rewarding, to combine the best parts of each of the candidate architectures to capture the possible synergy in a new or combined architecture. In the next section of these Conclusions we have pointed out the high points of each architecture studied. In our Recommendations section we have proposed one way by which the combined architecture could be developed.
- Recommendations of the Task Force to continue work on the extension or amalgamation of the candidate architectures in the framework of a possible International Research Program can be found in the following chapter, also there the Task Force develops recommendations for its own future with a new scope and organization.

19.2 SPECIFIC CONCLUSIONS REGARDING EACH CANDIDATE ARCHITECTURE

The following specific Conclusions concerning each candidate architecture as developed by the work of the Task Force or derived from the work of other groups can be presented:

19.2.1 CIMOSA

- The AMICE Consortium which initially developed CIMOSA are to be highly commended for their early decision to be as formal as possible in their definition and description of all aspects of this architecture. This was done with the ultimate aim of achieving complete computer executability of all constructs, models, tools, techniques, etc, associated with the architecture.

- The AMICE Consortium initially restricted the scope of CIMOSA to the discrete manufacturing field and in addition imposed a further restriction in the form of its application only to those factories where each element of the subject plant's production floor equipment has its own local control system. That is, CIMOSA studies only what is termed supervisory control and production management by other groups. The dynamic direct control of plant floor production units is handled only by the production unit's own built-in local controllers which are fed general operating instructions by the CIMOSA system.
- CIMOSA does have and describes a 'life history' for the CIM system. However, it has not extended this description to become a true methodology for the use of this architecture by the user organization's planning and development group for carrying out plant integration studies and programs.
- Because of its declared goal of formality and eventual computer executability, CIMOSA is the most formally described of the three candidate architectures. This has unfortunately imposed a burden on its readability and understandability particularly for those potential users who do not have a computer science or related educational background.
- In order to increase the potential for executability, CIMOSA is developing two major environments, the Enterprise Engineering Environment and the Enterprise Operation Environment and a set of specified system services known as the Integrating Infrastructure. The first of these environments formalizes the development of enterprise models and their conversion to working programs for the system. The second formalizes the testing, proving and acceptance of the resulting programs as new additions or changes to the operating systems. The Infrastructure defines how all such programs, as just noted, work together to carry out the overall functions of the integrated computer system.
- CIMOSA is the most thoroughly studied of the candidate architectures and the most widely published. It is also the one which has received the most attention concerning potential standardization as an International Standard.

19.2.2 GRAI–GIM

- GRAI-GIM has a well developed and described methodology for the application of its architecture and related tools and technology for the development of an integration program by the user through the specification stages.
- GRAI-GIM in its descriptive documentation has generally been confined to discussions of the development of the computer system and related hardware and software to implement the factory integration desired. Such a restriction is not inherent in the methodology or necessarily intended by the developers.
- Although not specifically so stated, GRAI-GIM discussions, descriptions, case studies, etc. are all to date confined to the discrete manufacturing field.
- GRAI-GIM has also confined itself to date in its life-cycle methodology to the concept, analysis, specification and detailed design phases of the overall development and use of an integration system for an enterprise. There is no discussion as yet of the construction, testing and commissioning of a system or of its continued development while in use. There is nothing inherent in the GRAI-GIM system itself which would prevent its ready expansion to cover these additional phases or indeed to study other industries than discrete manufacturing.
- Like CIMOSA, GRAI-GIM treats the human worker as a resource in terms of the required skills and physical capability. Also like CIMOSA there is no discussion of human relations, needed training programs, union requirements, human organization details, etc.
- The GRAI Laboratory in producing GRAI-GIM has developed several tools and techniques of potential wide use in enterprise integration studies such as the GRAI-GRID, GRAINET, ECOGRAI, the GRAI Model and others. They have also advanced the use of other tools and techniques developed by others such as MERISE, Petri Nets, etc.
- The philosophy of CIM and of the carrying out of a CIM development project as expressed by the GRAI-GIM descriptive documents is the best of three candidate architectures in terms of its adherence to the philosophy adopted by the Task Force and used in this Report.
- GRAI-GIM is intermediate between CIMOSA and Purdue in terms of the degree of formality implied and used and the consequent ease of understanding by the noncomputer science educated user.

19.2.3 PURDUE

- The Purdue Enterprise Reference Architecture and its associated Methodology are an informally described means for leading a user's application group through all of the phases of an enterprise integration program from initial concept through use to final plant obsolescence.
- As an informal description it is the easiest to understand by non-computer science educated users. This is especially true because of its easy to grasp graphical presentation of its overall structure and layered phases for program development. Its associated methodology is also the most complete, particularly in its discussion of the planning phases of an integration program. The organization of the Purdue Architecture is also important in that its structure approximates the way many plant workers think of their factories and their operations.
- The split of the Implementation View of the Purdue Architecture into Information System, Human and Organizational, and Manufacturing Equipment (or Customer Service) considerations permits the Purdue Methodology and the subsequent enterprise integration programs to undertake an extensive discussion of all of the human aspects of the enterprise involved as they affect enterprise integration.
- Likewise the manufacturing (or customer service) equipment considerations permit the details of the integration of material and energy flows, equipment layout optimization and many other important factors in the overall enterprise integration system to be considered beyond those of information integration alone.
- Purdue has shown its extendibility to cover all types of industries and indeed to cover all types of enterprises regardless of their individual mission and as a methodology is the most extensively documented of the three in its Purdue Implementation Procedures Manual.
- Purdue lacks the set of mathematical modelling techniques which is necessary when an architecture is to be computer modelled. There is no built-in restriction to adding these, so the architecture can be readily extended in this direction.

20

Recommendations for future work

Task Force

20.1 GENERAL RECOMMENDATIONS

Our general recommendations are as follows:

1. A funded Task Force program should be initiated which will carry out a case study on the three candidate architectures. A specific enterprise should be chosen (or defined) after which each of the architectural development organizations (of the candidate architectures) will be asked to provide an architectural description of that enterprise in terms of their specific architecture. Such study would be necessary for a detailed validation of the conclusions as drawn by the Task Force and give further insight into the applicability of the three candidate architectures.
2. The study by the Task Force initially concluded that, while neither of the candidate architectures was yet complete in terms of our definition of ultimate usability, there was no indication of unsurmountable difficulty in completing any of them if the initial development group or some follow-on group should decide to carry out this work. To reiterate, our definition of completeness includes the following:
   (a) In addition to modelling the proposed computer-based, scheduling and control system needed for information integration (under whatever title it is described), the architecture should also model the structure and progress of the project or program under which the task of integration of the enterprise, factory, or unit involved is achieved.
   (b) Along with showing the structural model of the project or program required, the architecture should have associated with it a methodology whose description will serve to instruct the user organization's planning and implementation groups in achieving the integration of information and of the flow of materials and products desired.

(c) In addition to instruction, the methodology should include aids (computer based or otherwise) which can be prescribed, or developed to whatever degree possible, to assist the user's planning and implementation group in their task of achieving the integration desired.

(d) The pathway should be clear to achieving a formally defined syntax and semantics for the models, techniques and tools involved in the architecture and its associated methodology. This should then allow the development of a computer executable form of the architecture compatible with current 'open systems' trends in computer usage.

Items 2a-2d need to be confirmed by the results of the Case Study of Item 1 above.

This report recommends a possible pathway by which this might be achieved for each of the candidate architectures and for a possible 'joint' one. Such work should be carried out for at least one complete architecture.

We further recommend that the above work should, if possible, be carried out in the framework of an International Research Program and it is to be determined how to plan, finance and implement such a program.

3. It is our observation that user groups in attempting enterprise integration feel overwhelmed by the mass of detail necessary to plan and carry out the integration project or program desired. Because of the size and complexity of the enterprises attacked in these studies the actual complexity involved is probably unavoidable. Therefore, the necessity which should be satisfied is to reduce the 'apparent' complexity involved. Some ways by which the above might be accomplished are:

(a) Development of instructional manuals of the methodology, and wherever possible, on the architecture itself, which are easily understandable by those not formally trained in computer science, and related technologies.

(b) Publication of comprehensive case studies using the architecture and methodology while thoroughly explaining each step of the process and its results achieved.

(c) Continued development of the possibilities and techniques of 'configurating' generic structures for scheduling and control systems, for communications systems, and for data base systems to ease the need to reprogram all or part of these systems for each new application.

(d) Develop as many as possible example populated system structures (generic or partial) which could be parameterized (and cropped or extended) as necessary by potential user groups. Again, this will help to avoid the repetition of the necessary collection of the massive detail involved. A compendium of such models already available would be most useful to develop.

(e) Search for techniques that reduce the apparent complexity of the architectures as experienced in the application context. The difficulty of dealing with the amount of data generated in realistic applications prompted researchers to look for simplification techniques of presentation and analysis to reduce apparent complexity, e.g. Precedence Matrix techniques [1]. Similar techniques could contribute to the ease of use of the reference architectures studied.

20.2 FURTHER DEVELOPMENT OF EACH ARCHITECTURE AND ITS ASSOCIATED METHODOLOGY

The Case Study recommended earlier should specifically address the following points concerning each of the three candidate architectures:

20.2.1 CIMOSA

In view of the findings of the Task Force and the general recommendations just above, it is suggested that CIMOSA could be developed into our so-called complete architecture as follows:

1. Make it clear in the description of the use of the CIMOSA Modelling Framework that it also applies to studies of the manufacturing equipment systems and of the human and organizational systems of the enterprise as well as to the information integration systems (the computer based systems).
2. Prepare a companion document to the existing descriptions of the CIMOSA Architecture itself to encapsulate the extensive discussion of the use of the CIMOSA 'life cycle'. This would provide the methodology to aid user groups who plan to employ CIMOSA for their enterprise integrating planning and development projects or programs. This documented methodology should also describe the needs and methodology of the identification and concept layers present in the other two architectures but missing, in description at least, in CIMOSA.
3. Remove the self imposed restriction of CIMOSA to discrete manufacturing systems. While all examples to date only treat such manufacturing systems, the Task Force could find no inherent limitations in CIMOSA that would prevent its expansion to include systems in other industries or, indeed, in enterprises in general.

20.2.2 GRAI-GIM

Like CIMOSA, GRAI-GIM is unnecessarily restricted, in the opinion of the Task Force, to the discrete manufacturing field. The following are other suggestions for 'completing' GRAI-GIM in the ways developed by the Task Force:

- GRAI-GIM is limited in its discussion to the manufacturing equipment and the control system. This is a self imposed limitation of treatment since the methodology to handle the human relations aspects could be readily included.
- Likewise the means for including the details of the modelling and applications of enterprise system construction, checkout, commissioning, etc. could be readily added. Similarly the technology for continuous improvement of the operations methodology applied to the working factory throughout its operational life could also be added for a complete life history treatment.

20.2.3 PURDUE

As stated in the Conclusions and Findings Section of this Report, the Purdue Enterprise Reference Architecture and its associated Methodology are the most informal of the three candidate architectures in their presentation. In view of the strongly expressed need for formality in order to achieve eventual executability of the architectural constructs, this appears to be the first need in further development of the Purdue offerings. This can be accomplished in several ways either separately or as a joint undertaking. Some of these are:

- Where applicable, employ the methodologies already developed by CIMOSA and GRAI-GIM, with appropriate credit as to source. Possible examples are the Integrating Infrastructure of CIMOSA and the GRAI GRID of GRAI-GIM.
- Where the required technologies do not already exist, in the companion architectures or other fields, develop them as needed.

Other developments which would be helpful to Purdue would be as follows:

- Purdue does not as yet require specific designated tools or techniques at each step of the development of an integration system, as called for by CIMOSA. It would greatly aid the user if work were carried out to classify, evaluate and document the relative applicability of each of the multitude of computer based, graphical and interactive analysis tools available.

This applies to every stage of the analysis, design, construction, checkout, and operation of the integration systems. Those which are especially applicable could be formalized, if not already accomplished, for that application.

- There is currently a major lack of formality in the specification and development of human and organizational systems. This involves the specification of the innovative tasks which must be carried out by humans, as well as the specifications of the appropriate organizations to promote the accomplishments of human based tasks. Also it should include the best ways of assuring the availability of the necessary human skills as needed. Anything which can be accomplished in this area would be extremely valuable for future integration projects.
- Formalization of the corresponding tasks in the manufacturing or customer service equipment area on the other hand is quite straightforward in concept. The problem here is the extremely wide range of devices, etc., which are needed and thus the corresponding massive endeavor needed to impose formality on this area.

20.3 A COMBINATION OF EXISTING ARCHITECTURES

In view of the strong points of each architecture as brought out in the several evaluations described herein and in the preceding Findings and Recommendations, the following is one possible consolidation of the three studied architectures which might be proposed. The Case Study proposed earlier should provide insight as to the correctness of these proposals:

- Use the structure of the Purdue Enterprise Reference Architecture to guide the overall program since it seems to be the one which is most easily understood and accepted by non computer science educated personnel in user planning and development groups.
- Use the CIMOSA Modelling Framework to supplement the Purdue data flow and material and energy flow analysis method. These include the generic task lists described there. These are necessary for establishing the functional requirements for the proposed overall system.
- Use the GRAI-GIM Modelling Framework as a reference check to assure the completeness of the requirements lists developed just above.

Complementing (or combinations of) existing methods and tools:

- In the Functional Design or Specification Level, apply the GRAI-GIM, GRAI GRID and GRAI NET as applicable and adopt the formality of the

CIMOSA Integrating Infrastructure, Integrated Engineering Environment and Integrated Operational Environment as templates for the design.
- Formalization of the Human and Organizational and of the Manufacturing or Customer Service Equipment areas for the Purdue Architecture.
- Adopt either the CIMOSA Integrating Infrastructure or the Purdue Manifestation and Operation Layers or Phases and complete them with the necessary formalization.

20.4 FUNDING AND COOPERATION

It must be stressed that none of the proposals and recommendations listed above can be accomplished unless the following also occurs:
- These proposals and recommendations are supported by the development organizations of the candidate architectures in terms of making the requested changes in their respective architectures or compromising as necessary to accomplish the requested proposals or recommendations,
- The contemplated work must not conflict with the ongoing International Standardization efforts where these apply, and
- Sufficient funds can be made available to have a small dedicated group to work on these proposals (and particularly the Case Study) to assure their completion.

20.5 A POSSIBLE FUTURE ROADMAP FOR PROMOTION OF THE FIELD OF ENTERPRISE INTEGRATION USEFUL FOR FUNDING AGENCIES

The work of the IFAC/IFIP Task Force on Architectures for Integrating Manufacturing Activities and Enterprises, as recorded in this Report, has shown that the needs of the users for their acceptance and application of their Integration Architecture in priority order are as follows:
1. A 'complete' Architecture is necessary.
2. An associated Methodology for applying the Architecture and its related models, techniques and tools in the easiest possible manner for User Application Groups must be available.
3. Standardization of the Architectures and their associated generic models, techniques and tools is very important as a promoter of their wide acceptability and use. It should be emphasized that it is the wide acceptability rather than standardization per se which is the important criteria here.

4. There is a general lack of knowledge of and appreciation of the capabilities of integrating systems to improve the productivity and economic return of enterprises of all types in the minds of potential users everywhere.

   This is due to the explosion of progress in computer and information systems technology in all areas of application in recent years and to the inability of the user community to keep abreast of these burgeoning capabilities and the associated benefits of their use.

   Thus any methods (case studies, training programs, easier to learn and use application methods, additional computerised tools, etc.) which promote knowledge and acceptance in the field are vitally important.

The Task Force believes that its current evaluation work has outlined what is necessary to achieve a 'complete' architecture. Likewise we believe that the requirements for a suitable methodology have also been defined, here and in the available literature. What remains is to do the work to achieve them.

Thus the Task Force would strongly encourage current and future funding organizations to promote funding policies which would help to achieve the needs just listed in this Section.

20.6 A PROPOSAL FOR FUTURE WORK FOR THE TASK FORCE

The major distinguishing feature of the Task Force in its work in evaluating and promoting the field of architectures for enterprise integration has been the fact that its membership included representatives of each of the major groups active in this field. Thus we are in a unique position to report on and hopefully guide the future development of this vitally important field for the future. Therefore, any future continuation of the Task Force and its work should continue to build on and exploit this unique characteristic and capability.

Since the Task Force believes its work during this past triennium has effectively answered the challenge of its initial Scope as presented by our organizers, a new Scope and Plan of Work is needed. Our recommendations for these are presented in the next section.

Should IFAC and/or IFIP desire that this work be continued along the lines proposed in the Scope and Plan of Work just proposed or a similar one, the Task Force and its members would be happy to oblige.

20.7 A PROPOSED SCOPE FOR ANY DESIRED CONTINUATION OF THE IFAC/IFIP TASK FORCE ON ARCITECTURES FOR INTEGRATING MANUFACTURING ACTIVITIES AND ENTERPRISES

20.7.1 Tasks to be undertaken

1. Monitoring of the initiation, development and execution of the Case Study and other development work proposed herein. It is presumed that this work will be carried out by personnel other then the Task Force members because of the degree of involvement and dedication necessary for their accomplishment.
2. Work toward the development of one or more complete enterprise integration reference architecture(s). This can be accomplished by (a) fleshing out either of the current candidate architectures in a manner similar to that proposed in the 1993 Report of the Task Force, or (b) by consolidating them into a best candidate similar to another proposal also made by the Task force in its report.
3. Evaluate the relationships of the currently available analysis, design and implementation models, techniques and tools and recommend a selection of these to be incorporated with the complete architecture(s) proposed.
4. Where appropriate recommend standards for models, techniques, and/or tools which will help promote the acceptance of the architecture(s) and associated methodolog(ies).
5. Develop an ongoing list of needed but primarily unavailable models, techniques and tools which would greatly aid the field of enterprise integration and encourage potential developer organizations to undertake this work.
6. Prepare syllabuses for innovative training programs needed in the area of enterprise integration and the details of integrating systems. As there are already many available programs for training in this area, only new, innovative ideas are worth promoting by an organization like the Task Force.
7. The Task Force would benefit from the inclusion of more members from the direct user community of enterprise reference architectures.

20.7.2 Method of operation

The technique of a Split Meeting used in the 1992 Spring Meeting to achieve a higher overall attendance from the membership did not fulfill its aim. Each of the Split Meetings was effectively another regular meeting period. Thus we achieved a larger number of meetings of a few active members. This resulted in a higher work product but did not result in attracting a larger number of the individual enrolled members to a particular meeting. We did learn some meeting organization techniques from our experience to date:

1. If possible there should be two meetings per year to allow time for a worthwhile work program.
2. These meetings should be distributed about the globe to aid members' travel plans. They should also be scheduled well in advance since many members have very long approval periods for foreign travel.
3. Now that the members know each other much better than before the initial formation of the Task Force it should be possible to carry out much more work by correspondence between formal meetings thus improving the work output.
4. Where possible, formal meetings should be scheduled the weekend ahead of established IFAC or IFIP Conferences or Workshops on relevant topics to our scope. This will aid the approval process for travel by many members.
5. Meetings should be 3–4 days in length to allow work proposals to develop worthwhile outputs while the group is together.

20.8 REFERENCES

[1] Eppinger, S.D., Whitney, D.E., Gebels, D.A. (1992) 'Organizing the Tasks in Complex Design Projects: Development of Tools to Represent Design Procedures,' 1992 NSF Design and Manufacturing System Conference, Atlanta, Georgia, January 1992.

Appendix A

The background of the Task Force

1.1 BACKGROUND OF THE IFAC/IFIP TASK FORCE ON ARCHITECTURES FOR INTEGRATING MANUFACTURING ACTIVITIES AND ENTERPRISES

The Task Force was formed at the 11th International Congress of IFAC (The International Federation of Automatic Control) in Tallinn, Estonia in August 1990 and confirmed by IFIP (The International Federation for Information Processing) at a meeting in Helsinki, Finland, a week later. In IFAC the Task Force was under the aegis of the Manufacturing Technology and Computers Technical Committees. In IFIP the cognizant Technical Committee was TC–5, Computer Applications in Technology.

From the initial IFAC proposal for the formation of the Task Force the Scope detailed in Section A.1.3 was developed by the Task Force at its first meeting in Bordeaux, France along with the corresponding set of proposed deliverables and the immediate work program included there. Members of the Task Force (Appendix B) are experts in the fields of Manufacturing Methods, Computer Integrated Manufacturing, Process Dynamics and Control, Mathematical Modelling, and Computer Science. They come from Europe, Japan, China, Canada, Australia and the United States.

Without in any way denigrating the excellent interests and contributions of all members of the Task Force, there were a few members who truly deserve special mention because of the outstanding contribution they have made to the work of the Task Force during this triennium. Without their contributions the progress recorded in this report would have been unlikely if not completely impossible. See Acknowledgements.

1.2 SOME OBSERVATIONS MADE AT THE EARLIEST WORKSHOP MEETING OF IMPORTANCE TO THE TASK FORCE'S FUTURE PROGRAM

The following observations were agreed to by the Task Force members at our earliest meeting as characterizing the Enterprise Reference Modelling Field at that time:

- There are a very large number of models already in the literature or developed as proprietary projects by many industrial groups.
- None of these are complete as yet.
- Most present many of the same concepts but by means of different graphical and mathematical methods.
- The ancient parable of the group of blind Indian philosophers who attempted to describe an elephant after each had felt only different separate parts, certainly applies here – each of the proposed architectures is describing the same subject but from widely varied, and very incomplete viewpoints. Thus the descriptions appear to be very different.
- The work of the Task Force is to sort all of this out – as noted above. Our first attempt will be through a comparative analysis of the available models. We will then attempt to distill the specifications for the best model from the existing proposals.
- The best solution would be to be able to adopt one existing model or architecture as best – failing this, build on an existing system by supplying the apparently missing capabilities – the last resort would be to specify a whole new one from the characteristics of several of the others.

The effect of these observations on our scope and work program is very evident from the following section 1.3.

1.3 SCOPE AND WORK PROGRAM FOR THE IFAC/IFIP TASK FORCE ON ARCHITECTURES FOR INTEGRATING MANUFACTURING ACTIVITIES AND ENTERPRISE

- Exploration of Suitable Architectures Including CIMOSA and Other Open–System Architectures.
- Investigation of Methods to Use the Architecture to Design a CIM System Including Computerized Tools to Support the Methods.
- Exploration of Aspects of Manufacturing Knowledge at Management, Support and Operational Levels for Design, Planning, Production, and

Installation and Service Tasks in the Integrated Manufacturing system Environment.

- Establishment of the Information Systems Paradigms that Support its Architectures.
- Investigation of Protocol and Interface Requirements.
- Determination of the Relationships of the Architectures to the New Generation of Controllers for Cells and Enterprises.
- Determination of the Generic Aspects of the Architecture(s) Important to Other Types of Control Systems

1.4 SUGGESTED PRODUCTS OF THE TASK FORCE

- Make a Comparison of the Architectures Studied by Models Used, Views Presented, Etc.
- Identify (Point Out) the Usefulness of the Architecture(s) to Provide an Analysis Context for Enterprises.
- Attempt to Collapse the Work Under Item 1 Above Into a Single Architecture Applicable to all Industries. At Least Identify the Generic Items and Issues Involved.
- If Possible Create a List of Appropriate Research Protocols and an Associated Road Map of Possible Interest to Funding Agencies.

1.5 IMMEDIATE WORK PROGRAM

There is a vital need for a reference architecture which can model all aspects of the CIM system development cycle and of the life history of the CIM project from its original concept in the eyes of management through its full life cycle to final plant obsolescence. This work program proposes to further the development of such a reference architecture by carrying out the following steps:

- Making a comparative analysis of all of the major existing enterprise reference models and architectures. This analysis will be made through a comparative evaluation of the breadth and completeness of the coverage of each and a mutual mapping of the features of each upon the others along with other suitable comparison methods which may be developed in the future.
- Based on the above analysis we propose to select a best such architecture by:

(a) adopting one of the existing reference architectures as clearly the best; or

(b) failing this, by completing the best of the existing ones by supplying those parts found missing by the above analysis; or finally

(c) if neither of the above is practical due to incompleteness of all members of the trial set, then to develop a new one using the best features of those examined (with due credit to the sources, of course) along with the needed features missing from all.

1.6 REQUIREMENTS FOR A REFERENCE ARCHITECTURE TO BE USED AS A SET OF EVALUATION CRITERIA

An enterprise reference architecture should provide the study group contemplating a new CIM project or other enterprise endeavor with the following capabilities:

- A structure or pattern outlining all the steps and requirements for the proposed endeavor thus assuring completeness in the analysis and development of the project at hand.
- Examples of requirements which must be fulfilled at each stage of the development of the project thus easing the design and specification task involved.
- A characterization of the analysis and design tools, whether graphical, computer based or manual available to ease the workload at each step of the design and development project.
- A listing of pertinent standards applicable for analysis, design or operation of the facility under consideration.

See also Sections 3.3 and Table 3.2.

Appendix B

Membership list of the IFIP/ IFAC Task Force

IFAC/IFIP TASK FORCE ON ARCHITECTURES FOR INTEGRATING MANUFACTURING ACTIVITIES AND ENTERPRISES

Active members (positions and addresses as of 1993)

Dr. Peter Bernus
School of Computing and
Information Technology
Griffith University, Nathan, (Brisbane)
Queensland 4111, Australia

Mr. Jim Brosvic, Consultant
Advanced Automation Center, 2S19
Honeywell, Inc.
16404 N. Black Canyon Highway
Phoenix, AZ 85023 USA

Dr. David Chen
Laboratoire GRAI
Universite de Bordeaux 1
351 cours de la Liberation
33405 Talence CEDEX, France

Professor Guy Doumeingts, Director
Laboratoire GRAI
Universite de Bordeaux 1
351 cours de la Liberation
33405 Talence CEDEX, France

Mr. Fadi G. Fadel
Department of Industrial Engineering
Rosebrugh Building
4 Taddle Creek Road
University of Toronto
Toronto, Ontario
Canada M5S 1A4

Dr. Atsushi Inamoto, Manager
Manufacturing Industry Department
Information and Communication
Systems Engineering Center B,
Mitsubishi Electric Corporation,
Kongoh Building 7–10–4,
Nishigotanda Shinagawa–ku,
Tokyo 141, Japan

Dr. Laszlo Nemes, Executive Officer
Strategic Science and Technology
Division of Manufacturing Technology
CSIRO, Locked Bag No. 9
Preston, Victoria 3072, Australia
Chairman – IFAC Manufacturing Technology Committee

Dr. Gustav J. Olling
Automotive Process Research & Technical Planning Executive Engineering Office, Chrysler Motors Corporation
12000 Chrysler Drive CIMS 418–0–4–18
Highland Park, MI 48228–1118 USA
Resigned – January 1992

Mr. James L. Nevins
26 Sunset Drive
Burlington, MA 001803, USA

Dr. Theodore J. Williams – Chairman
Professor of Engineering and Director
Purdue Laboratory for Applied Industrial Control, Purdue University
1293 Potter Engineering Center
West Lafayette, IN 47907–1293 USA

Dr. Bruno Vallespir
CIM Research Project Leader
Laboratoire GRAI
Universite de Bordeaux 1
351 cours de la Liberation
33405 Talence CEDEX
France

Ing. Marco Tomljanovich
ITALCAD Tecnologic e Sistemi S.P.A.
Via Panama, 52
00198 Rome
Italy
Chairman – IFIP TC–5, Computer Applications in Technology

Ing. Jakob Vlietstra, Author
1942 Buckeye Road
Willits, CA 95490

Ing. Dick Zoetekouw
AT&T Network Systems Nederland BV
Post Office Box 1168
1200 BD Hilversum
The Netherlands
Resigned – October 1992

Corresponding members (positions and addresses as of 1993)

Mr. Randy Aranguren
IBM Corporation
D202/B658A, Highway 52 & Northwest 37th Street
Rochester, MN 55901 USA
Convenor, ISO TC 184/5C 5/WG 1

Dr. Albert Benveniste
Director de Recherche de INRIA
IRISA–INRIA, Campasde Beaulieu
35042 Rennes CEDEX, France
Chairman – IEEE CS – IFAC TC Project

Professor Mark S. Fox
Department of Industrial Engineering
University of Toronto
4 Taddle Creed Road
Toronto MSS lA4, Canada

Dr. Zengjin Han
Professor of Control Engineering
Department of Automation
Tsinghua University
Beijing, Peoples Republic of China

Professor Leo Motus
Chairman IFAC Computers Committee
Head of Division, Institute of Cybernetics, Estonian Academy of Sciences
Lenini Puiestee 10, Tallinn 200104
Estonia; Chairman – IFAC Computers Committee

Dr. G. Kovacs
Computer and Automation Research Institute, Hungarian Academy of Sciences, P. 0. Box 63
H–1518 Budapest Hungary

Professor Karl–Johan Astrom
Dept. of Automatic Control
Lund Institute of Technology
P. 0. Box 118
5–221 00 Lund
Sweden

Professor Chen Yuliu
Department of Automation
Tsinghua University
Beijing 100084
Peoples Republic of China

Dr. Yoshiro Fukuda
Manager of Production System Section
Technical Research Institute, JSPMI
12 Hachiman–cho, Higashikurume–shi
Tokyo, 203, Japan

Professor Lennart Ljung
Department of Electrical Engineering
Linkoping University,
S–581 83 Linkoping, Sweden
Chairman – IFAC Technical Board

Professor Michael G. Rodd
Department of Electrical and Electronic Engineering
University College of Swansea
Singleton Park, Swansea SA2 8PP, UK
Past Chairman – IFAC Computer Committee, Member – IFAC Technical Board

Index

GPSR Compliance

The European Union's (EU) General Product Safety Regulation (GPSR) is a set of rules that requires consumer products to be safe and our obligations to ensure this.

If you have any concerns about our products, you can contact us on ProductSafety@springernature.com

In case Publisher is established outside the EU, the EU authorized representative is:

Springer Nature Customer Service Center GmbH
Europaplatz 3
69115 Heidelberg, Germany

Batch number: 10761588

Printed by Printforce, the Netherlands